Occasional Paper No 20

Scientific Studies in Early Mining and Extractive Metallurgy

edited by
P. T. Craddock

Elln MacLame

British Museum
1980

BRITISH MUSEUM OCCASIONAL PAPERS

British Museum, Great Russell Street,
London, WC1B 3DG.

Executive Editor : G.B. Morris, BA
Production Editor : Ms K.E. Symonds, BA

Occasional Paper No. 20, 1980 : Scientific
Studies in Early Mining and Extractive
Metallurgy.
P T Craddock.

ISBN 0 86159 019 8

ISSN 0142 4815

Cheques and postal orders should be made
payable to the Trustees of the British Museum
and sent to the Accounts Office at the above
address.

CONTENTS

INTRODUCTION

This volume forms part of the proceedings of the 19th International Symposium on Archaeometry and Archaeological Prospection which was jointly organised by the British Museum Research Laboratory and the Institute of Archaeology of London University and held in London in March 1979.

The symposium formed one of a series which were organised in Oxford until 1975 but which have had a different international venue in Europe and North America each year since then. The first time that the proceedings were published, however, was following the 18th Symposium held in Bonn. In 1979, it was decided not to publish all the papers together, but to group them into four volumes according to subject matter (British Museum Occasional Papers Nos. 18-21). This was the first Symposium to devote a session exclusively to Early Mining and Extractive Metallurgy and it's inclusion is symptomatic of the dramatic increase in interest and activity in the subject during the past decade.

Archaeology is the study of man's material past with all the potentials and limitations attendant on the study of ancient sites and the artifacts from them. Our knowledge of man's prehistoric social and economic development can at best only be inferred from such indirect source material. However the technology of the artifacts, of what and how they were made can be much more precisely studied. The firm parameters of the material information allow deductions in this field which are far less speculative, being directly related to the artifact and it's place of production. Prehistoric archaeology may never be able to explain why a given votive offering was made but from it's scientific and archaeological study together with that of a related production site it should be possible to unequivocally determine how it was produced. There is a lot to recommend pursuing the arts of the possible!

Most previous studies on ancient technology have concentrated on the study of the surviving artifacts themselves. This is reasonable as the numbers of artifacts hugely outweighs the number of production sites available for study, and they tend to be concentrated in museum and university collections conveniently near to scholarly and scientific institutions. Much can be learnt from these studies, especially of the later stages of fabrication. One has only to cite the example of the excellent Freer Chinese Bronzes publication where a whole volume was devoted to a detailed description by Gettens of the technology of the bronzes, although even here reference was made to the finds from the ancient casting sites at An Yang and Hou-Ma. However this approach still leaves open the question of how the ores were mined and smelted to produce the metals from which the artifacts were subsequently made. When the technological investigation moves back to the mining and pyrotechnology stages one needs the archaeological information from the production centres themselves and ideally the investigating scientists should be available on site during the excavation to record and sample in concert with the equally important standard archaeological recording. This is another practical reason why many studies on early metal production have had to be content with vague descriptions of unexcavated early mines and chance finds, coupled with speculations on the smelting process based on surviving primitive practices, and common sense inferences from metallurgical chemistry, and with a study of the relevant artifacts. Unfortunately, although eminently useful, surviving primitive techniques display a bewildering variety of potentially comparable possibilities from which to choose, and the metallurgical chemistry of early processes is not so well understood as one would have hoped.

Interest is recording primitive metallurgical processes dates back at least to the second half of the 16th century with the De Re Metallica (1556) of Georg Bauer, called Agricola and Sir Walter Ralegh's The discoverie of the large, rich and beautiful Empire of Guiana of 1596. However, it was not until the 19th century that deliberate investigation of the remains of early mining and metallurgy was commenced. William Borlease in Cornwall, the Siret brothers in Iberia, and Ardaillon at Laurion in Greece, made notably local investigations on the ground and the century culminated in the outstanding work of William Gowland whose studies encompassed recording primitive metallurgy in the Far East, the metallographic study of antiquities and the actual investigation of early metalworking sites. His intergrated approach, tackling the problem of ancient metallurgy using anthropology, archaeology and science foreshadows modern methods and makes his copious literature and ideas especially valuable now. This broad approach spanning the disciplines is well reflected in the position he held, Vice-President of the Society of Antiquaries of London, and President of both the Royal Anthropological Institute and of the Institute of Metals, the only time these posts have been held by the same man. In this century Davies produced his detailed survey of Roman mines throughout the Empire and after World War II Pittioni began his detailed scientific and archaeo-logical investigation of the Alpine Bronze Age copper industry which was the first major modern survey of ancient mining and metallurgy to be carried out. At the end of the 1950's Rothenberg began his work in the Wadi Timna, Israel, followed shortly afterwards by Jovanovic at Rudna Glava, Jugoslavia; both are continuing projects and some of the latest research associated with these sites is published in this volume.

The more purely scientific investigation of ancient metallurgy based on the study of artifacts began with the work of Klaproth, the father of Analytical chemistry in the late 18th century, and continued with the excellent work of von Bibra and Bluemner in the 19th. During a large part of this century scientific and especially analytical work on metals was often channelled away from technological studies to those of typology. In this approach, the scientific description of the artifact, and especially of its trace element composition was used as part of the typological description in an attempt to better order, and ultimately provenance the artifacts, and to identify the source of the copper. Some of the papers in this volume show the difficulties encountered and why a simple analytical approach was bound to fail. This concentration upon the accidental, rather than deliberate features of the metal contributed little to the history of technology, and ultimately, little to typology either.

Since World War II the more technologically orientated research of metallurgists and scientists such as Marechal, Salin, Barnard, Coghlan and Tylecote have built up an impressive body of information, which has complimented well the growing body of archaeological information from mines and smelters discussed above. The past decade has seen the integration of and the close cooperation between physical science (especially chemistry and metallurgy) and archaeology resulting in the rapid expansion of knowledge throughout the field of ancient metallurgy that we see today.

Archaeology has always had to call on a very wide range of specialist help from other disciplines and this is especially true of the study of ancient mining and metal-lurgy. With the growth of archaeological excavations of ever increasing sophistica-tion and the funding of a wide range of back up research facilities, some of the former outside specialist work can now be covered by workers concentrating full-time on the archaeological aspects of their specialist knowledge and skills. The success and growth of environmental studies in archaeology is perhaps the best established example of this, but the physical sciences are now playing an ever in-creasing part in the study of the past, especially of it's technology, and this involve-ment is clearly essential in the study of ancient mining and metallurgy. The recent interest of institutions such as the British Museum Research Laboratory and the Deutches Bergbau Museum, Bochum in ancient mining and metallurgy, and the establishment of the Institute of Archaeo-metallurgical Studies at the Institute of Archaeology, London are welcome indicators of the long term cooperation of the physical sciences and archaeology.

The interpretation of contemporary early descriptions of ancient processes where they exist is of course still extremely important. Berthelot's magnificent work at the close of the 19th century on all aspects of early science, provided the foundation for the 20th century work notably by Caley, Sherwood Taylor, Forbes,

Needham and C.S. Smith.

The final digested product of the archaeological, anthropological, scientific and textual work must always remain the experimental reconstruction and operation of the metallurgical process carried out to check the hypothetical reconstruction and to add yet more information on the process.

This stage can only be attempted after sufficient parameters governing the various processes have been established to allow a reasonably unequivocal reconstruction to be made. At present in ancient metallurgy this is only possible with the furnace remains from a very few sites since on the majority there are still important gaps in our knowledge.

Unfortunately modern mines often sit athwart the ancient mines and with ever increasing wholesale and efficient mining and extractive processes the ancient deposits are being swept away including the remains of some of the most famous mines of antiquity. In practice little can be done to preserve these sites, and we must accept that they will be destroyed. However we can ensure that excavation is carried out in advance of destruction, and in this the modern mining companies have been most cooperative, not only allowing the excavations to proceed but also giving logistical and often financial aid as well, coupled with a welcome and keen interest in the research shown by all personnel from truck driver to managing director. This policy of rescue archaeology in front of destruction can very often mean having to record and collect refractory, slag, and other material whose significance cannot now be fully assessed. It will be the task of future workers to piece together the processes from what we can record and preserve now. It is thus vitally important that adequate and detailed study collections of technological remains and debris are made and preserved, not only in the local museum adjacent to the particular sites together with the more orthodox materials from the excavations, but also in major international museums where comparative collections can be built up near to the advanced research facilities and institutions whose resources now and in the future will be required to fully elucidate the processes represented by the archaeological remains. Fortunately the debris from mining and smelting is very prolific and substantial collections can be made without detracting at all from the main archaeological collection. The slags and furnace remains may seem unprepossessing, but it is vital that such collections are made and properly stored as they will be the key to future work and progress in the subject. Soon many sites will have gone and only be represented by what we can now save. The future must interpret the past through our eyes.

The introduction and development of metalworking has been described as one of the main forces behind craft specialisation, technological innovation, and even of urbanisation. Much has already been written of the products of the smith in antiquity, now we are beginning to understand the processes by which they were made.

BRONZE AGE COPPER MINING IN COUNTIES CORK AND KERRY, IRELAND

JOHN S. JACKSON

Department of Geology, Trinity College, Dublin

Abstract

Seventy four mines of proven or probable Early Bronze Age date in southwest Ireland, (west Co. Cork and southwest Co. Kerry), are described or noted. Seventy three are of 'Mount Gabriel type', drift mines of modest dimensions opened on greenish-grey sandstones of Old Red Sandstone (Devonian) age and containing syngenetic copper. One mine at Derrycarhoon, of considerably greater dimensions, exploited a discordant vein-type deposit. Copper concentration (overall average) runs at 0.5 precent, (5000 p.p.m.) in the syngenetic ore. The total tonnage of rock, gangue and ore mined in the area has been estimated at 77,782 tonnes with an estimated copper metal content of 414 tonnes. A production of 372.5 tonnes of finished copper metal is based on an assumed smelter conversion efficiency of 90 percent. The total weight of copper and bronze artifacts cast during the Early Bronze Age in Ireland has been estimated as some 75 tonnes of metal. When this is compared with the estimated total of finished copper metal produced in southwest Ireland alone a substantial imbalance becomes apparent, the ratio being 1:5. This imbalance points inescapably to the conclusion that Ireland was an important nett exporter of copper during the Early Bronze Age. Wind-rose diagrams from south and southwest Ireland indicate the southern and western slopes of Mount Gabriel as most probable smelting pit areas, diametrically located to the main mining area on the mountain.

Keywords: MINING, COPPER, IRELAND, BRONZE AGE, METALLURGY, COPPER PRODUCTION, GEOLOGY, MINERALOGY.

INTRODUCTION

In a recent publication the production of copper ore in Ireland during the Bronze Age was provisionally estimated as equivalent to some 1464 tonnes of copper metal, (Jackson, 1979). It was noted that "it is accepted that such an estimate must of necessity be approximate but it would at least reflect the general scale of copper mining in the country at that time and it is therefore considered of interest to attempt such an approximation. As additional information becomes available from other areas this preliminary estimate can be revised and refined". (op. cit.) Additional data are now available and it is therefore considered justifiable to attempt a revision of the earlier estimate but, as explained below, only for the southwestern part of the country.

In those parts of Ireland in which medieval and later mining took place the evidence of prehistoric mining has, in large measure, been obliterated. For example, Ptolemy refers to Oboca (Avoca), a copper mining area intermittently worked since prehistoric times and included on his map of Ireland of about 150 A.D. The mines at Barrystown and Clonmines in County Wexford were worked prior to the

Normans in the 1170s and in 1557, in the reign of Philip and Mary, there is a comprehensive reference to "the setting forth of the King and Queen's Majesties' mines in Barristoune by Claymyne, in the County of Waxfforthe..." (Carew MSS.). The mines in the Silvermines area of County Tipperary were worked in the 13th century and in the year 1289 ten pounds "of the King's treasure (was) paid to William de Cerne to defray expenses regarding the King's Mine in the County of Tipperary." (Gleeson, 1937). These mines were developed and worked by members of a colony of Florentine and Genoese merchants, a situation comparable to the Macclesfield merchants who took over and developed the Avoca mines at the end of the 18th century. And there are many other examples of medieval and later mining in the east of Ireland and in the Irish midlands.

In the Irish midlands the bogs are of the raised type, (<u>hochmoor</u> of Germany), in which peat growth was initiated several millenia before the Early Bronze Age. Therefore the Bronze Age miners had perforce to prospect for copper in those areas outside the boglands and their mines would therefore have been exposed and accessible to later miners to be re-worked and modified and unequivocal evidence of earlier workings obliterated. In the southwest of Ireland, however, the blanket bogs developed considerably later, after the exploitation of the copper mines, and here the mines were concealed by peat growth and their existence only realised when peat was cut away for fuel in the 19th and 20th centuries. The Mount Gabriel mines, for example, were unknown to geological surveyors in the 1850s and were first noted in the late 1920s after the local peat cover had been removed. Also the type of copper mineralisation in the southwest of the country favoured the development of small drift-mines in which the ore grade was low and these mines were therefore of limited interest to the larger mining companies of the 18th and 19th centuries and, although they were cleared out and inspected, many of them survived virtually intact, e.g., the drift-mines at Derrycarhoon, Horse Island, etc.

In the southwest of Ireland many of the Bronze Age mines are as the miners of that period left them and even where subsequent re-working has taken place it was of comparatively recent date and there is therefore often a record in the literature testifying to their great antiquity, for example, Ross Ireland, Derrycarhoon, Horse Island, Ballyrisode, etc. A much more reliable estimate of Bronze Age copper production can therefore be attempted for the southwest region than from any other area in Ireland. Because of these factors the area of Bronze Age mining and copper production has, in this paper, been confined to the southwest of the country, including west Co. Cork and southwest Co. Kerry.

COPPER MINERALISATION IN SOUTHWEST IRELAND

Copper mineralisation in the south and southwest of Ireland is distinctive and quite unlike that in the rest of the country. In order to appreciate this fully it is necessary to look briefly at the geology of the area and the palaeo-environmental conditions under which the sediments and the associated copper were deposited. In 1971 preliminary results of a field study "of disseminated copper mineralisation occurring within the sedimentary rocks in an area of about 500 square miles ($1295km^2$) in the southwest corner of Co. Cork" were presented, (Snodin, 1971): this included a statigraphic succession for the rocks of southwest Cork, as follows:

Table 1

	Stratigraphic Formation	Stratigraphic Thickness	Palaeoenvironment	Copper mineralisation
Lower Carboniferous (Tournaisian)	Black Slate Formation	4000 ft. (1219.2m)	Marine	None
	Coomhola Formation	2000 ft. (609.6m)	Deltaic	Copper mineralisation confined to lowest part of succession.
Upper Devonian	West Cork Sandstone Formation (containing the Sherkin Member.)	6000 ft. (1828.8m)	Coastal flood-plain with meandering rivers.	Widespread copper mineralisation but most concentrated in the uppermost part of the succession. Mineralisation very rare in the Sherkin Member.

As can be seen from Table 1 the copper mineralisation is most concentrated in the uppermost part of the West Cork Sandstone (Upper Devonian) and near the base of the overlying Coomhola Formation (Lower Carboniferous). The copper was noted to occur preferentially in beds of greenish grey or grey colour and to be "concentrated within the coarser grained lithologies of the greenish grey strata, that is within the river channel sandstones and cornstones of the West Cork Sandstone Formation and within the deltaic sandstones of the Coomhola Formation..." The copper is syngenetic, i.e., was laid down with the enclosing sandstones, and was presumably detrital. Due to mineralisation being essentially confined to the sandstones in-filling channels of meandering rivers incised into a coastal flood-plain, the thickness of any single copper-bearing bed is limited, usually to 1-2m. The concentration of copper within these beds can range up to two percent but the overall average is some 5000 p.p.m., i.e., 0.5 percent. (Discussion of Snodin, 1971).

The shallow water environment associated with copper deposition was one of coastal flood-plains (initially) and, with deepening water and the onset of marine conditions, deltas. These features were associated with and controlled by the margin of a geosyncline, or subsiding basin of sedimentation, which lay across the South of Ireland. This geosynclinal margin corresponds essentially with a feature referred to in the literature as the Variscan Thrust Front, (see the broken line marked VTF, in figure 2a), (Gill, 1962). Current geological field studies question the thrust front interpretation and at present this feature is regarded as of structural significance, controlling sedimentation, rather than being of tectonic origin. In the context of Bronze Age copper mining this line is of great importance for it represents the northern limit of the copper mineralisation associated with the greenish grey sandstone and siltstone beds and is therefore the northern limit of Bronze Age mines of the Mount Gabriel type. The apparent maverick mine on Ross Island, near Killarney, (figure 2a), which lies to the north of the VTF line, is in a completely different type of mineralisation emplaced in the Carboniferous limestone and is clearly epigenetic, i.e., emplaced at a time considerably later than the deposition of the enclosing limestones.

Figure 1a (inset).
Map of Ireland
showing, in black,
area described in
text.

Figure 1b.
Map of Southwest
Ireland showing
study area in
relation to the
cities of Cork and
Limerick.

Figure 2 (inset).
Area shown on
enlarged scale in
Figure 2a.

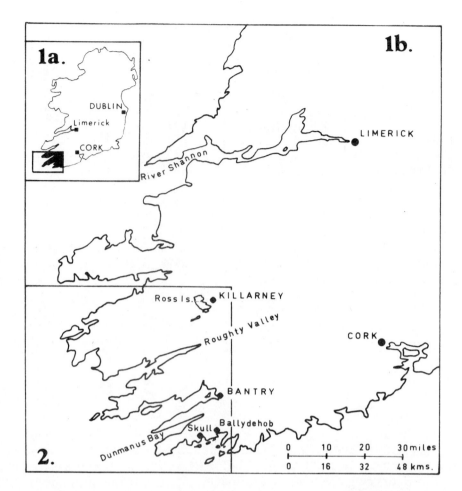

DISTRIBUTION OF BRONZE AGE COPPER MINES IN SOUTHWEST IRELAND

The Bronze Age copper mines occur on three peninsulas, the Mizen Peninsula, the Beara Peninsula and the Ivernian Peninsula, and also in the Killarney area. These will be described in some detail below. The location of each site is shown in figure 2a.

(3) (1) County Cork.

(3) (1) (i) Mizen Peninsula.

(a) Mount Gabriel

A number of additional mines on Mount Gabriel have been noted since the complex of Early Bronze Age copper mines was described in 1968 when twenty five mines, (Nos. 1-23 with two additional mines 3a and 3b) were presented in map form, (Jackson, 1968, figure 2). In 1970 the Commissioners of Public Works in Ireland made a National Monuments Preservation Order, (No. 2 of 1970 dated 26th March, 1970), covering the Bronze Age copper mines and their tip heaps on Mount Gabriel. During the survey of these mines, specifically undertaken for the purpose of the Preservation Order by engineering staff of the Office of Public Works, four additional mines were recognised, now referred to as Mines 7A, 13A, 21A and 24, and brought the total to twenty nine. These are shown in figure 3. Recently Dr. Tom Reilly, Geological Survey of Ireland, has completed a detailed geological field mapping programme in west Co. Cork, a revision of an earlier survey, during which he located two additional mines on the southwestern slopes of Mount Gabriel, so that the total noted to date is thirty one. The mines are distributed through four town-lands, as follows: Mount Gabriel td. (18); Letter td., (3); Rathcoole td., (1);

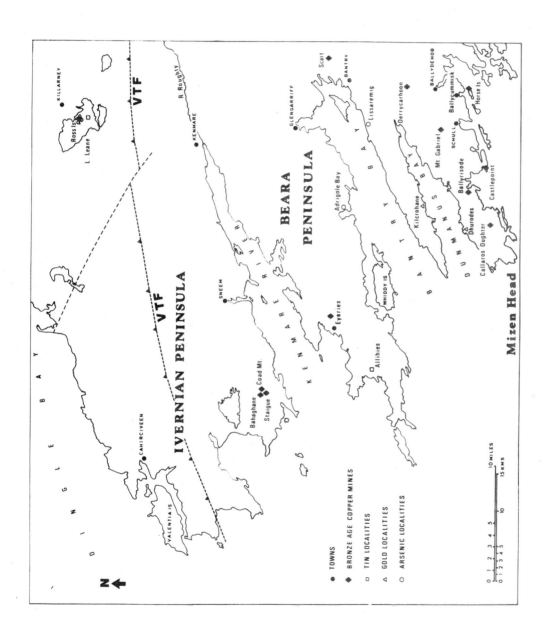

Figure 2a: Map of west Cork and southwest Kerry showing Bronze Age copper mines together with tin, gold and arsenic localities. VTF = Variscan Thrust Front. (The Cork/Kerry county boundary passes down the Beara Peninsula; Glengarriff and Eyries are in Co. Cork and Kenmare in Co. Kerry).

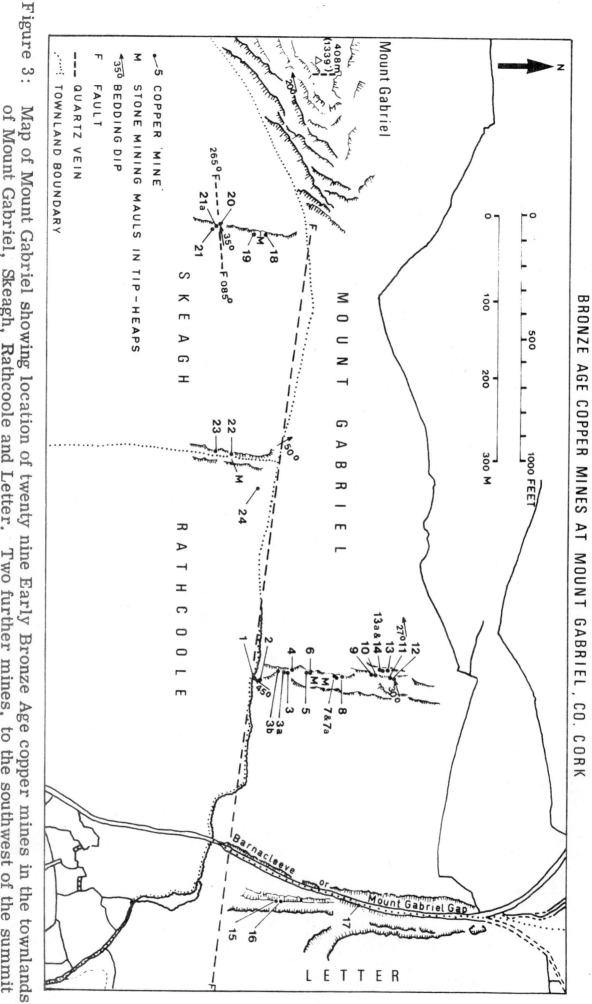

BRONZE AGE COPPER MINES AT MOUNT GABRIEL, CO. CORK

Figure 3: Map of Mount Gabriel showing location of twenty nine Early Bronze Age copper mines in the townlands of Mount Gabriel, Skeagh, Rathcoole and Letter. Two further mines, to the southwest of the summit of Mount Gabriel, lie outside the area of the map.

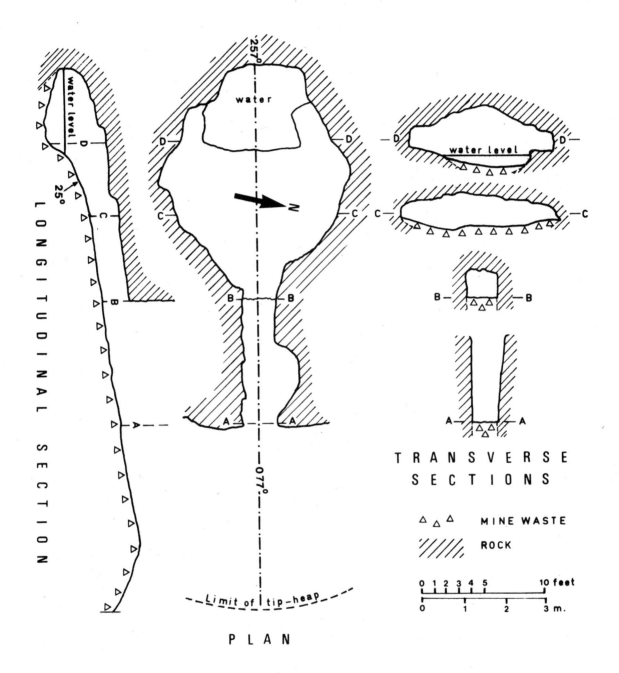

Figure 4: Mount Gabriel. Mine plan of Mine No. 1. (Modified from Jackson, (1968), Figure 3, p. 104).

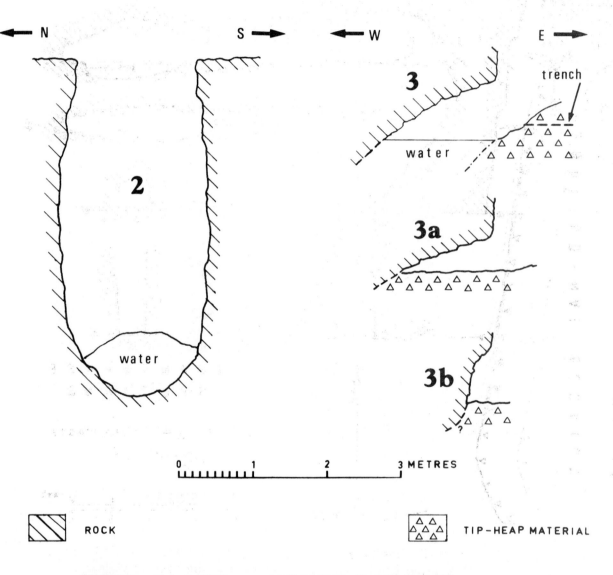

N ←   S →   ← W   E →

3

trench

2

water

water

3a

3b

?

0   1   2   3 METRES

ROCK   TIP–HEAP MATERIAL

MINE PLANS — MOUNT GABRIEL

Figure 5: Mount Gabriel. Mine plan of Mine No. 2.

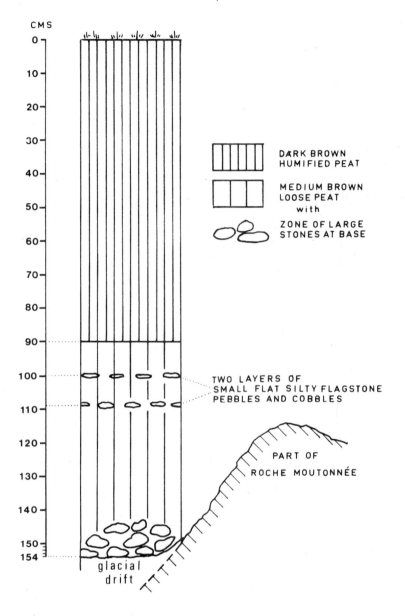

PEAT SUCCESSION — 91.4 m. N.E. OF MINE NO.5. MOUNT GABRIEL.

Figure 6:   Mount Gabriel. Blanket bog profile in surviving peat area, 91.4m
            northeast of Mine No.5. The large stones at the base of the profile
            were probably carried down by hill-wash from the adjacent mine areas
            at the beginning of the wet period which initiated the growth of the
            peat.

Skeagh td. , (9). As the Mount Gabriel mines have already been described in some
detail elsewhere, (Jackson, 1968; 1979), no further description will be given here.
Mine plans for two of the mines are given; Mine No.1 is presented in figure 4 and
Mine No.2, not previously published, is illustrated in figure 5. Reference has
already been made to blanket bogs sealing off the copper mines in the general area
under discussion. On Mount Gabriel residual patches of peat survive here and
there. The profile of such a surviving patch, some 1.5m thick, occurs to the
northeast of Mine No.5. and is illustrated in figure 6. The large stones at the

base of the profile are interpreted, in part, as waste from the adjacent mine tips. The peat of the western type blanket bogs does not lend itself to dating by palynological techniques and the peat has not yet been radiometrically dated; it clearly post-dates the mines. The undisturbed thickness is estimated to have been in excess of 4.27m, i.e., that recorded at Derrycarhoon.

(b) Derrycarhoon

The early copper mines at Derrycarhoon were first discovered by Capt. Charles Thomas and have been described by Kinahan, (1885), Windele, (1862) and Jackson, (1968; 1979). They were mapped on a scale of 1;2500 by Duffy, (1932). They will not therefore be described in detail here. Two types of mineralisation occur: (i) "...a mineral vein emplaced discordantly within the Old Red Sandstone beds with a length of 18.288m by 18.288m deep and a width of 1.219m" and (ii) "six drivings similar to those on Mount Gabriel." (Jackson, 1979). These are presented in figure 7.

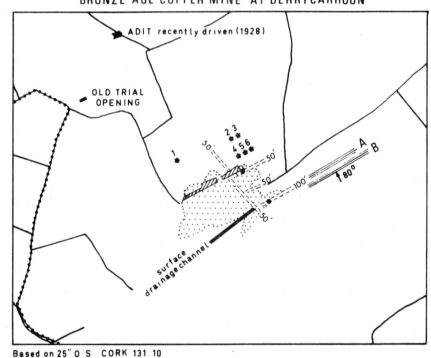

BRONZE AGE COPPER MINE AT DERRYCARHOON

Based on 25" O S CORK 131 10

•   DRIFTS or SHALLOW TRIALS (? BRONZE AGE)

●   SHAFTS

======   DRIVINGS (at 50´or 100´levels)

TIP HEAP (including Bronze Age material)

BRONZE AGE COPPER MINE

Figure 7:   Derrycarhoon, north of Ballydehob, showing major Bronze Age copper mine (lined area) and seven 'Mount Gabriel type' drivings, (1-7), with 19th century shafts and drivings. The extensive tip heap (stippled) comprises Bronze Age mine waste (including broken grooved mining mauls) together with mine waste of the 19th century operation. Based on MS map prepared by T. J. Duffy, (1932). Scale 1: 2500.

(c) Horse Island, Roaringwater Bay

Although the copper lodes on Horse Island in Roaringwater Bay were re-worked in
the first half of the nineteenth century and much of the previous workings obliterated,
there is considerable evidence in the literature to demonstrate compellingly that
copper mining took place there during the Bronze Age. In a deposition to the
Landed Estate Courts, (1855), dealing with the Estate of Lord Audley, County of
Cork, Ireland, there is a "statement as to the Copper Mines, Slate Quarries, etc.
in Lot 10", (i.e., part of the Audley Estate). Referring to copper, (p. 11), the
deposition of Samuel Jacob states that in March 1835 he "visited the Audley Mines
for the purpose of opening the works at Horse Island, and upon that occasion he was
accompanied by Mr. John Munday of Cornwall... that they commenced operations
on Horse Island; and in a few days, in the eastern side of the said island, sunk a
shaft and raised up some tons of rich Purple Sulphuret of Copper, yielding from
20 to 60 percent of copper. He also upon that occasion employed to the westward
of that shaft, ten or twelve men, who in ten days, merely by spades and shovels,
took off the surface of the cliff about 12 tons of the carbonate of copper, yielding
about five percent of copper: that in the month of October, 1835, the two shafts
in the Eastern extremity of Horse Island, under the direction of Mr. Munday, were
then producing a considerable quantity of rich copper ore, at the rate of about 3 tons
per week..."
    The copper mineralisation on Horse Island is, as on Mount Gabriel, restricted
to greyish-green sandstone/siltstone beds. "Malachite, in small quantity, is
usually seen at the surface... azurite and chalcopyrite are also occasionally seen
at the surface... In depth... the ore seems to have existed in the form of irregular
lenses and pockets." (Duffy, 1932). This type of mineralisation has been postulated for Mount Gabriel where it would have controlled the distribution of the
drivings. (Jackson, 1968).
    It is apparent that Samuel Jacob and John Munday also cleared out and developed
pre-existing drivings on Horse Island for Adam Murray, in a report to Lord Audley
dated November, 1833, remarked that on Horse Island he "saw a great quantity of
hazel nuts among the old workings in the northeast part of the island." This information is of particular relevance and importance. No hazel, in fact virtually
no trees, now grow on the island. Charcoal of hazel (Corylus) was recovered with
that of oak (Quercus) and willow or poplar (Salix/Populus) from the tip-heap outside Mine No. 5 on Mount Gabriel, a mountain now cloaked in blanket bog and
virtually devoid of trees, (Jackson, 1968). This demonstrated that hazel grew in
the area prior to the deterioration of the climate which initiated the development
of blanket bog on the Mizen Peninsula; hazel represented 21.4 percent of the charcoal examined from the tip of Mine No. 5. The base of the blanket bog, (see
figure 6), has not been radiometrically dated but there is evidence to suggest that
peat growth started on the Mizen Peninsula at an early date, probably quite soon
after the development of the mines on Mount Gabriel. (See Sect. IX (c) in Jackson,
1968, p. 100). This killed off the arboreal vegetation of the area, including the
hazel, and it can therefore be deduced a priori from the occurrence of hazel nuts
in the old workings on Horse Island that these mines are of Bronze Age date.
    The Ordnance Survey surveyed County Cork in 1841-2, (Andrews, 1974,
figure 4, p. 19), some eight years after Murray's report, and six inch Sheet No. 149
(Cork) shows five 'copper mines' in the vicinity of Parkcooleenapeasta, to the
southwest of Cus Point, and 'Karkeet shaft' and another mine near the coast to the
east of Carrigleegarode. These refer in part to the recently re-opened ancient
mines and to new drivings, e.g., 'Karkeet shaft', but the map also shows a
'Danish shaft' to the west of Parkcooleenapeasta. This term refers quite clearly
to ancient workings and would have been based on local oral folk tradition; at Mount
Gabriel the Bronze Age drivings are referred to locally as 'Danes' Mines' or 'Old
men's workings' and the term 'Danes Island' refers to an isthmus in Co. Waterford
with a concentrated group of Bronze Age mines, (Jackson, 1979, figures 1 and 2).

Further evidence in the literature for a Bronze Age date for the Horse Island drivings is given by an anonymous Cork correspondent to the Mining Journal. In 1864 he noted that about forty years ago the West Cork Mining Co. cleared out old Danes' shafts containing "curious implements", presumably stone mining mauls, etc., (Anon, 1864). Dr. T. Reilly of the Geological Survey of Ireland considers that the correspondent's memory was at fault and that the period to which he referred was about thirty years prior to publication, i.e., about 1834, and not 1824 as the above account indicates.

It is fortunate that this evidence is available from the literature for little now survives on Horse Island to unequivocally demonstrate the existence of mines of Bronze Age date. The question has to be answered regarding the number of Bronze Age mines which existed on Horse Island. One mine, the 'Danes shaft' noted as such on the Ordnance Survey map, clearly refers to an ancient working and the anonymous correspondent's observations clearly demonstrate that there was more than one such driving when he refers to "old Danes' shafts containing curious implements". Murray also refers to a number of mines when he notes that he "saw a great quantity of hazel nuts among the old workings..." From the style of the drivings I have interpreted six of the drivings on Horse Island as belonging to the Bronze Age. Because of the relatively early growth of the blanket bogs which has been postulated for the general area it can be inferred that the pre-historic mines of west Cork belong to the Early Bronze Age.

(d) Castlepoint

Duffy (1932) notes this locality and observes that "about 240 yards (219.5m) E.N.E. of the old castle there is an old shaft...locally called the "Old Men's working." This term, in common with 'Dane's shafts', is based on local oral tradition; the Mount Gabriel mines are referred to locally as 'Danes' Mines' or 'Old Men's workings'. There is therefore an a priori basis for regarding the Castlepoint shaft as of antiquity and, although lacking unequivocal evidence, possible of Bronze Age date.

(e) Ballyrisode (=Ballyrizzard).

Ten or twelve mines, similar to those on Mount Gabriel, have been described by Caulfield, (1880), and associated polished stone axes have been described and figured and their dating significance discussed, (Jackson, 1968).

(f) Callaros Oughter

Duffy, (1932), notes that "an old shaft known as the Dane's shaft" occurs here. It is apparently shallow and was sunk on a "bed of (grey) grits...in places stained with malachite." This description, including depth, lithology and copper staining, is reminiscent of the Mount Gabriel drifts and the local tradition -- 'Dane's shaft' -- strongly indicates antiquity.

(g) Scart

This locality, in the vicinity of Bantry at the head of the Mizen Peninsula, was noted by Dr. Tom Reilly during his survey of west Co. Cork. He noted an old driving similar to those on Mount Gabriel and I am indebted to him for bringing this hitherto unknown mine to my attention.

(3) (1) (ii) Beara Peninsula

(a) Eyries

Mines of Mount Gabriel type and with associated stone mining mauls occur on the ground above Eyries, northeast of the copper mine at Allihies, the latter worked in the 19th century, (Somerville-Large, 1972). Somerville-Large refers to "mines" with associated stone mining mauls. (op. cit. , p. 205) but does not state the number of mines. Only one is therefore quoted in Tables 2 and 3 below.

(3) (2) County Kerry

(3) (2) (i) Ivernian Peninsula

(a) Behaghane

This locality, on the west slopes of Coad Mountain, is noted by Duffy, (1932), who states that an opening known locally as St. Crohane's Hermitage "was apparently an old opening made on the copper-carbonate stained quartz reef in a search for copper -- probably in prehistoric times." Duffy was familiar with the Mount Gabriel and Derrycarhoon mines and in offering an opinion that the Behaghane opening was "prehistoric" he would have been making a carefully considered statement.

(b) Staigue

In describing this locality Duffy, (1932), notes that, about 550 yards (503m) S. S. E. of Staigue Fort, (a dry stone masonry ring fort probably dating to the Iron Age), "there is a bed of coarse grey grit with thin strings of malachite. Traces of old workings -- possibly pre-historic -- are seen in this bed of grits." He does not state the number of these workings but there are obviously more than one.

(c) Coad Mountain

A shallow driving on greenish grey sandstones occurs on Coad Mountain. Its general appearance coupled with the lithology of the beds are reminiscent of the Mount Gabriel drivings.

(3) (2) (ii) Killarney Area

(a) Ross Island, Lough Leane, Killarney, Co. Kerry

The antiquity of the copper mines on Ross Island, an isthmus on Lough Leane, Killarney, Co. Kerry can be inferred from the prodigious quantity of grooved stone mining mauls which occur on the lake shore in the vicinity of the mines. There is a reference to several hampers being filled with stone mining mauls during a picnic on the island in the early 19th century, for subsequent distribution to friends, (Hall, 1841,), and the mauls have been collected intermittently from this site for over a century. Notwithstanding this, in 1965, during a visit to the site, I collected thirty one reasonably well preserved examples of these mauls, each exhibiting the 'equatorial groove', now in the collections of the National Museum of Ireland; register nos: NMI. G1 : 1965/1-31. (These are tabulated in Jackson, 1968).

Unfortunately the ancient workings were totally obliterated by mining activities in the 19th and early 20th centuries but fortunately there is a record in the literature of the re-opening and re-working of these mines in the early 19th century which confirms their antiquity. Writing in 1824, T. Crofton Croker observed that "the remote antiquity of Ross Mine is established by a discovery made on clearing out the old shafts when it was re-opened a few years since, at which time several rude implements of stone were found buried under decayed vegetable matter and rubbish... Marks of fires were distinctly to be seen when the rock was exposed to view, which, with the discovery of the stone implements, affords satisfactory evidence of the mine having been originally worked at a period prior to the knowledge of either iron or gunpowder, and hence local tradition attributed these operations to the Danes." (In parenthesis in should be noted that "the term 'Danes' does not necessarily refer to a Viking" (Jackson, 1968), or to an invader from Denmark but rather to mining activities in the remote past.)

The reference to fire-setting and to the occurrence of stone mining mauls embedded in an accumulation of decomposed organic 'rubbish' could be applied, with accuracy, to the situation in the mines on Mount Gabriel and Derrycarhoon, which are known to be of Early Bronze Age date.

I have estimated, on the evidence of the phenomenal quantity of grooved stone mining mauls already collected from the site and still surviving on the foreshore of the lake, that the equivalent of ten mines of Mount Gabriel type were driven during the Bronze Age on Ross Island and that this estimate is probably conservative. (Jackson, 1979, Table 2). Unlike Mount Gabriel, copper mineralisation on Ross Island is emplaced in the Dinantian (Lower Carboniferous) limestone, brecciated by faulting, and the style of mineralisation is therefore quite different as also, presumably, was the type of mine driven.

(4) Estimated number of Bronze Age copper mines in southwest Ireland

From the description of mine sites presented in Section (3) above an estimate of the total number of those of proven or probable Bronze Age date can be made.

Table 2

Bronze Age copper mines in southwest Ireland

County	Area	Locality	Number of copper mines
Cork	Mizen Peninsula	(1) Mount Gabriel.	31
		(2) Derrycarhoon.	(6 drivings (1 major mine
		(3) Horse Island.	6*
		(4) Castlepoint.	1
		(5) Ballyrisode (=Ballyrizzard)	12
		(6) Callaros Oughter..	1
		(7) Scart	1
	Beara Peninsula	(8) Eyries	1
Kerry	Ivernian Peninsula	(9) Behaghane	1
		(10) Staigue	2*
		(11) Coad Mountain	1
	Killarney Area	(12) Ross Island.	10*
		Total number of mines ...	73 + 1 major mine

(* = estimated. See text of Section (3) for rationale of estimates).

As can be seen from Table 2, there is an estimated total of 73 mines of Mount Gabriel type in southwest Ireland plus one major mine at Derrycarhoon. This estimate is considered conservative. For example, Duffy, (1932), described copper occurrences in 148 townlands in Counties Cork and Kerry, excluding those in the Allihies mining district of Co. Cork. He records modern trials, shafts, adits, opencasts and other operations from 85 of this number, representing 57.4 percent of the total. Many are short drivings on the greenish-grey copper-bearing sandstones, suggestive of re-worked ancient mines but no field evidence or local oral tradition survives to either prove or suggest their antiquity. None of these has therefore been included in the above total in Table 2.

(5) Method of estimating Bronze Age copper production

An estimate of copper (finished metal) production involves four parameters:
(a), the total number of mines, (b), the average size of the mines, (c), the concentration of copper within the ore mined and (d), the efficiency of converting the
copper in the ore, by roasting and smelting, to the finished metal.

(a) Number of mines

The number of mines operating during the Bronze Age in southwest Ireland has
been estimated as 73 which, for reasons already given, is considered conservative.
To this number is added a single major mine (at Derrycarhoon).

(b) Average size of the mines

Jackson, (1979), has given dimensions for three mines on Mount Gabriel and the
volume of rock, gangue and ore extracted from one of them, (Mine No. 4), was
estimated as $450.28m^3$, equivalent to 1198 tonnes, (conversion factor of 2.66
tonnes/m^3). The average mine plan area for the three mines was $18.1m^2$, (op.
cit., Table 1), and that for Mine No. 4 was $20.64m^2$. The average mine area,
expressed as a percentage of Mine No. 4, is therefore 87.7 percent. Assuming
that volume is directly proportional to mine area, the mass of rock, gangue and
ore extracted from the 'average' mine would be 1050.65 tonnes, or 87.7 percent
of that extracted from Mine No. 4. It is assumed, for estimation purposes, that
this average applies to the area as a whole. For the estimated total of 73 mines,
therefore, the total tonnage extracted would be 76697.45 tonnes, (equivalent to
$394.89m^3$ (average) x 73 = $28827m^3$ in volume), of rock, gangue and ore.

(c) Concentration of copper

The copper concentration in the sandstones intersected in the mines was previously
estimated at 2 percent, (Jackson, 1979). As has been shown in Section (2) above,
Snodin, (1971), quotes 0.5 percent as the overall average in these sandstones over
a vertical thickness of 1-2m., i.e., a thickness greater than the height from mine
floor to mine roof in the Mount Gabriel mines, e.g., 1.5m in Mine No. 1, (Jackson,
(1968). This value of 0.5 percent is therefore used in this paper as the overall
average copper concentration. Therefore the total tonnage of 76697.45 tonnes of
rock, gangue and ore extracted would contain 383.49 tonnes of copper metal.
   The single major mine at Derrycarhoon was driven on a discordant mineral
vein in which the concentration of copper would have been substantially greater
than the 0.5 percent syngenetic copper in the greenish-grey sandstones. Ten
percent by volume (copper mineral) was used in an earlier estimate, (Jackson,
1979); a more conservative estimate of 5 percent copper mineral by volume is used
here. The copper equivalent of the ore extracted from Derrycarhoon (major mine)
was previously estimated as 60.88 tonnes of copper metal; this is now revised to
30.44 tonnes. (This figure, calculated on a different basis to the straightforward
0.5% copper by weight, is derived as follows: dimensions of major mine = 18.288m x
18.288m x 1.219m = $407.69m^3$ x 2.66 = 1084.5 tonnes of rock/gangue/ore. At
5% by volume of copper mineral = $20.38m^3$ x 4.48 tonnes, (average density of
chalcopyrite/tetrahedrite/tennantite) = 91.32 tonnes @ 33.3% copper content = 30.44
tonnes copper metal).
   The total weight of copper metal mined in Southwest Ireland during the Bronze
Age would therefore be 413.93 tonnes, i.e., 383.49 + 30.44 tonnes.

(d) Smelting efficiency

Since no smelting pits or associated slag have been discovered as yet on any of the mining areas described above it is impossible to estimate the percentage of copper remaining in the slag. The efficiency of converting the copper in the ore to finished metal by roasting and smelting is therefore unknown. Previously a conversion factor of 50 percent was proposed, (Jackson, 1979), but this is now regarded as unacceptably low. The conversion efficiency depends in large part on the number of people available, and particularly women and children, for cobbing the ore and concentrating the copper in the feed to the smelter by removing gangue. I now consider a conversion factor of 90 percent as feasible. The 413.93 tonnes of copper metal in the ore could therefore have produced 372.5 tonnes of finished copper metal.

(6) Estimated copper production

The figures quoted above can be presented in tabular form, giving estimated production of rock, gangue and ore mined, copper metal content and finished metal, and the contribution made by each individual mining locality or district with sub-totals for each of the four major areas described.

Table 3

Mining locality or district	Number of mines	Rock, gangue and ore mined (tonnes)	Metal content (at 0.5% conc.) (tonnes)	Smelted finished metal, (at 90% efficiency.) (tonnes)
(1) Mount Gabriel	31	32570.15	162.85	146.56
(2) Derrycarhoon	6	6303.90	31.52	28.37
	1 (major)	1084.50	30.44	27.40
(3) Horse Island	6	6303.90	31.52	28.37
(4) Castlepoint	1	1050.65	5.25	4.73
(5) Ballyrisode (=Ballyrizzard)	12	12607.80	63.04	56.74
(6) Callaros Oughter	1	1050.65	5.52	4.73
(7) Scart	1	1050.65	5.52	4.73
Sub-totals for the Mizen Peninsula	58	60937.70	304.69	274.22
	1 (major)	1084.50	30.44	27.40
	58 + 1	62022.20	335.13	301.62
(8) Eyries	1	1050.65	5.25	4.73
Sub-totals for the Beara Peninsula	1	1050.65	5.25	4.73
(9) Behaghane	1	1050.65	5.25	4.73
(10) Staigue	2	2101.30	10.50	9.46
(11) Coad Mountain	1	1050.65	5.25	4.73

Sub-totals for the) Ivernian Peninsula)	4	4202.60	21.00	18.92

(12) Ross Island Lough Leane	10	10506.50	52.50	47.30

Sub-totals for the) Killarney Area)	10	10506.50	52.50	47.30

Total production for southwest Ireland.

Mizen Peninsula.....	58 + 1	62022.20	335.13	301.62
Beara Peninsula.....	1	1050.65	5.25	4.73
Ivernian Peninsula...	4	4202.60	21.00	18.92
Killarney Area	10	10506.50	52.53	47.28
Totals	73.+ 1	77781.95	413.91	372.55

Some interesting facts emerge from the tabulated results presented in Table 3. Of an estimated total 413.91 tonnes of copper mined, 335.13 tonnes (80.97 percent) came from the Mizen Peninsula and of this almost half, (48.59 percent) came from the Mount Gabriel mining district. The next area in importance is Ross Island on Lough Leane, Killarney which produced an estimated 52.53 tonnes (copper equivalent), or 12.69 percent of the estimated total production. The remarkably low figure for the Beara Peninsula, accounting for only 1.27 percent of the total, is of interest. Allihies, an important 19th century copper mining area, is located at the western end of this peninsula, southwest of Eyries, and was discovered at the beginning of the 19th century by the copious green staining of copper carbonate (malachite) on the outcropping rocks in the area. It seems remarkable that it could have been missed by Bronze Age miners and it is conceivable that the early mines were obliterated by the 19th century workings, which were substantial, and that no record was published at the time.

(7) Significance of estimated copper production in southwest Ireland

It has been noted that almost 2,500 "copper and bronze objects have been found in Ireland which date back to the Copper and Early Bronze Ages." (Harbison, 1966). The weights of ten randomly selected Early Bronze Age axeheads are presented in Table 4.

Table 4.

Register No; (National Museum of Ireland)	Description of artifact:	Locality:	Weight:
1932:6426.	Flat bronze axehead.	Clashaganny td., Co. Galway.	265 grams.
1933:5076.	Flat bronze axehead.	Cavan town district, Co. Cavan.	340 "
1936:1781.	Bronze axehead.	nr. Broughshane, Co. Antrim.	210 "

1937:82.	Flat bronze axehead.	Cotton td., Bangor Parish, Co. Down.	502	''
1939:33.	Flat bronze axehead.	Mortarstown Upper td., Co. Carlow.	465	''
1939:36.	Flat bronze axehead.	Dublin.	465	''
1939:50.	Flat bronze axehead.	nr. Cork.	160	''
1959:66.	Small bronze axehead.	'Crocknagrave School-house', Knockagrave td., Co. Monaghan.	58	''
1960:820.	Flat bronze axehead.	Deerpark td., Raphoe Co. Donegal.	198	''
1966:16.	Flat bronze axehead.	Cahirguillamore td. (or Rockbarton td.), Co. Limerick.	290	''

Total weight   2953 ''

Average weight (10 axes)   295.3''

(Maximum wt... 502 grams.
Minimum wt... 58 ''   )

The average weight of the axeheads tabulated is 295.3 grams. Using a rounded off value of 300 grams the 2,500 objects would weight some 750kg. Assuming that the 2,500 objects represent only one percent of those fabricated the total weight would come to 75 tonnes. This estimated total production of Early Bronze Age artifacts would apply to Ireland as a whole.

There is growing evidence pointing to the pre-historic copper mines of south-west Ireland being of Early Bronze Age. The Mount Gabriel mines date to 1500 ± 120 B.C., (Jackson, 1968; Felber, 1970). The thickness of the peat profile in the blanket bog which covered the Derrycarhoon mines suggests an initial period of growth in about 1600 B.C. and, although it is accepted that this figure is not precise, in general terms it suggests that peat growth could have been initiated in southwest Ireland prior to the Late Bronze Age, thus preventing access to these workings. If the copper mines of southwest Ireland are exclusively of Early Bronze Age date then a spectacular imbalance becomes apparent between the estimated weight of fabricated artifacts on the one hand, (75 tonnes), and the estimated weight of finished copper metal produced in southwest Ireland alone, (372.55 tonnes), or a ratio of 1:4.97, rounded off to 1:5. The conclusion appears inescapable that southwest Ireland, and probably Ireland as a whole produced copper far in excess of her needs in the Early Bronze Age and must have been an important nett exporter of the metal, a conclusion already noted elsewhere, (Jackson, 1979). In terms of the economics of trading and balance of payments, this would have permitted Early Bronze Age man to barter copper, probably in ingot form, for gold, tin and other requirements and might explain the remarkably high incidence of unfinished bronze axeheads and 'ingots', (Harbison, 1966, figure 3), and flat axe moulds, (op. cit., figure 4), in Co. Antrim, an area palpably poor in copper but rich in gold, (Jackson, 1979, Map 1). Perhaps copper was 'exported' from South-west Ireland in exchange for gold from northeast Ireland, a metal conspicuously rare in west and southwest Kerry.

(8) Location of smelter sites on Mount Gabriel

As already mentioned, no smelting pits or associated slag have been found as yet on Mount Gabriel. Most attention has been concentrated on the areas of mining

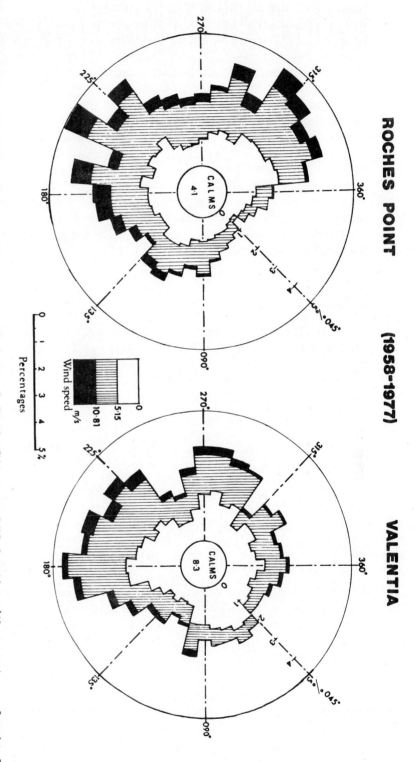

Figure 8 : Wind rose diagrams for Roches Point, southeast of Cork City, and Valentia, (the meteorological station is now on the mainland, southwest of Cahirciveen). For locations of meteorological stations see maps (figures 1b and 2a). Period covered is the twenty years from 1958 to 1977. Note the maxima for 210° and 310° (Roches Point) and 280° and 180° (Valentia). The azimuthal angular separation is 100° in both cases and the Roches Point maxima have been rotated clockwise through a 30° arc. Percentage frequency of wind directions shown on radial scale (at 045° in both diagrams). Wind speeds, expressed in metres per second, shown by progressive shading.
(Diagrams prepared from data supplied by the Meteorological Service, Department of Tourism and Transport, Dublin).

28

activity on the eastern slopes of the mountain. However, this may be mistaken.
Wind rose diagrams (figure 8) have been prepared from data from meteorlogical
recording stations at Roches Point, the headland forming the eastern side of the
entrance to Cork Harbour, (see figure 1b), and at Valentia, Co. Kerry, (see
figure 2a); the station is now located on the mainland, southwest of Cahirciveen.
It can be seen that two maxima occur in both stations, those at Roches Point being
210° and 310° while those at Valentia are 180° and 280°, the azimuthal angular
separation being 100° in both cases. From the geographical locations of these two
stations it can be assumed that prevailing winds at Mount Gabriel would be orientated approximately midway between those at Roches Point and Valentia, i.e.,
195° and 295°, and it is also assumed that there was no substantial difference between the direction of prevailing winds in Early Bronze Age times and the present.
It can also be seen in figure 8 that the highest incidence of strong winds, in excess
of 10.81 metres per second, coincide essentially with the wind direction maxima.
The obvious location for smelting pits would be on ground exposed to the strong
prevailing winds, that is a little west of south, (195°), and north of west, (295°),
on Mount Gabriel and not on the eastern flank, protected by the bulk of the mountain
from the effects of the winds. The search for smelting pits should therefore be
directed to the southern, southwestern and western flanks of the mountain, not on
the east as previously.

Acknowledgements

I am indebted to a number of colleagues for assistance during the preparation of
this paper. I am particularly grateful to Dr. Tom Reilly of the Geological Survey
of Ireland who made available to me a considerable body of information based on
his recent field work in west Co. Cork, during which he discovered three previously
unrecorded copper mines of Mount Gabriel type, and who brought to my attention
important information contained in the Landed Estate Courts papers and other
sources. His help has been invaluable.
My thanks are also due to Mr. J.J. McCarthy, Chairman, Office of Public
Works (Ireland) who kindly made available to me detailed maps of Mount Gabriel
prepared by O.P.W. technical staff in connection with a National Monuments
Preservation Order. Mr. Patrick Butler, Meteorological Service (Climate
Section), Department of Tourism and Transport, Dublin made available computer
print-outs of wind data from Roches Point and Valentia, the two nearest meteorological stations to the study area with continuous recording facilities.
Mr. Raghnall O Floinn, National Museum of Ireland, gave me data on the weights
of Early Bronze Age axeheads in the National Collections which are presented in
Table 4. My thanks are also due to Mrs. Diana Large who prepared most of the
figures and to Mr. Jimmy Cuffe, Department of Geology, Trinity College, Dublin
for the preparation of photographs. Finally it gives me particular pleasure to
thank Dr. Paul Craddock for his many kindnesses to me in his capacity of editor
of these Proceedings.

BIBLIOGRAPHY

Andrews, J.H.   1974.   "History in the Ordnance Survey Map -- an intro-
                                 duction for Irish readers". Ordnance Survey
                                 Office, Dublin. 63pp.

Anon.   1864.   "Account of mining in West County Cork".
                                 Mining Journal, 34, 559.

Carew MSS... Item 213.   "Mines" in Carew manuscripts preserved in the Archipiscopal Library at Lambeth, 1515-1574, (London, 1867, pp. 268-272). Item 213 also refers to "Mint", (213v).

Caulfield, R.   1880.   quoted in Journ. Roy. Soc. Antiq. Ireland, 15, 341-2.

Croker, T. C. Crofton.   1824.   "Researches in the South of Ireland, illustrative of the Scenery, architectural remains, and the manners and superstitions of the peasantry : with an appendix containing a private narrative of the Rebellion of 1798". London.

Duffy, T. J.   1932.   "Copper deposits in southwest Ireland". Unpublished typescript report in the Geological Survey of Ireland.

Felber, H.   1970.   "Vienna Radium Institute Radiocarbon Dates I", Radiocarbon 12, 1, 298-318.

Gill, W. D.   1962.   "The Variscan Fold Belt in Ireland", pp. 49-64 in "Some Aspects of the Variscan Fold Belt" K. Coe (ed.) Manchester Univ. Press. 163pp.

Gleeson, D. F.   1937.   "The Silvermines of Ormond" Journ. Roy. Soc. Antiq. Ireland, 67, 101-116.

Hall, S. C. and A. M.   1841-3.   "Ireland, its scenery, character, Etc". London. (3 vols).

Harbison, P.   1966.   "Mining and Metallurgy in Early Bronze Age Ireland," North Munster Antiq. Journ., 10, 3-11.

Jackson, J. S.   1968.   "Bronze Age copper mines on Mount Gabriel, west County Cork, Ireland" Archaeologia Austriaca, 43, 92-114.

Jackson, J. S.   1979.   "Metallic ores in Irish prehistory: copper and tin" in "The origins of Metallurgy in Atlantic Europe" Proceedings of 5th Atlantic Colloquium, Dublin, 1978. M. Ryan, ed.

Kinahan, G. H.   1885-1889.   "Economic Geology of Ireland", Journ. Roy. Soc. Ireland, 8 (n. s.), 1-514.

Landed Estate Courts.   1855.   Papers relating to the Estate of Lord Audley, County of Cork, Ireland. London.

Snodin, S. R.   1971.   "The nature and origin of copper-rich sedimentary rocks in south west Ireland" Cyclostyled abstract of paper presented at Symposium on "The genesis of base metal deposits in Ireland" Irish Geological Association; Galway, 1971. pp. 1-4, (abstracts separately paginated).

Sommerville-Large, P.   1972.   "The Coast of West Cork" London.

Windele, J.   1862.   "Ancient Irish gold and its origin.." Ulster Journ. Arch., 2, (1st ser.), (1861-62), 197-222.

Plate 1A: Group of six stone mining mauls from main tip-head (see figure 7) at Derrycarhoon, Co. Cork. All specimens show either abrasion or broken surfaces through use.

Plate 1B: Stone mining maul from Derrycarhoon, Co. Cork. Enlarged photograph of maul '1' in Plate 1A. 'G' = pecked groove; 'E' = abrasion at either end of specimen.
(Specimens collected by the author and now in the Collections of the Geological Museum, Trinity College, Dublin).

PRIMARY COPPER MINING AND THE PRODUCTION OF COPPER

BORISLAV JOVANOVIĆ

Institute of Archaeology, Knez Mihajlova 35 /II, IIooo Beograd, Yugoslavia

Abstract

Recent results obtained at Rudna Glava, northeast Yugoslavia, have shown the existence of the earliest copper mining, developed at the time of Vinca culture. This culture belongs to the Early Eneolithic of southeast Europe, with absolute dates of the first half or the middle of the fourth millenium B. C.

The data obtained at Rudna Glava support the proposition of considerable production of copper during the Early Eneolithic of southeast Europe.

Keywords: COPPER, MINING, VINČA CULTURE, EARLY ENEOLITHIC, BALKANS, YUGOSLAVIA, METALLURGY.

INTRODUCTION

The development of the oldest copper metallurgy in southeast Europe was based on discovery of local sources of raw material, through constant activity of contemporary copper mining. But mining was not invented for obtaining metals alone, because an advanced knowledge of minerals and their exploitation existed throughout the Neolithic period.   There are a lot of well known examples of the extraction of flint in Europe by means of the oldest mining technology. To put it more simply the use of flint was certainly older than the use of copper; hence the mining of flint must be older than the mining of copper. The first exploitation of copper deposits is not a new invention in the European Neolithic, but rather the application of an older technology for winning a new raw material. The two other stages of the production of the earliest metallurgy - smelting of copper ores and manufacturing of the metal objects - are much better investigated and determined.

Therefore one has the impression that the investigation of the copper metallurgy has been carried out back to front! The first investigations were directed to the oldest copper objects, as the final products of the primary metal industry. Those objects were typologically classified, and their composition and function studied. Further investigation of their manufacture led to the problem of the raw material. To turn to mining, it has been clear that the basic data did not exist. The oldest mines were not known, and the earliest commonly known, examples were dated to the Late Bronze Age, when the metal industry in Europe had attained a very high level. Investigations carried out in the Austrian Alps, for example, have shown that some mines, like Mitterberg, represented already developed mining centres, with massive production and a highly evolved mining technology (Pittioni, 1971).

Thus the beginning of mining was not quite clear, and alternative theories have been based on the intensive, local use of native copper, or on the diffusion of metallurgical experience from the Near East. According to the first hypothesis the amount of native copper in regions rich in ore deposits was quite sufficient for the beginning of copper metallurgy. Therefore large scale mining was not necessary during this initial phase. According to the second, diffusionist theory, one should expect rather developed mining technology in the Balkans, as it was in the neighbouring region of Anatolia. It is justifiable to expect that the spread of metallurgical and mining knowledge came from a much higher technological level; this is why the initial phase of copper mining southeast Europe was influenced by diffusion of techniques.

Recent investigations in South Bulgaria at Aibunar, (Chernych, 1978) and northeast Yugoslavia at Rudna Glava, (Jovanovic, 1978), have revealed, as is already known, the oldest mining sites is southeast Europe. Now it is possible to see the first direct examples of ancient mining in southeast Europe, and to check to what degree the knowledge of mining could help in determination of the contemporary metal industry.

All the facts, obtained so far, speak in favour of rather developed, independent, local mining, which corresponded to the territories of the different cultural groups of the Early Eneolithic of the central and south Balkans. This conclusion is based on the analyses of the material culture from the two sites, Ai Bunar in Thrace and Rudna Glava in north - east Serbia. It also means that the extraction of copper ores in the Early Eneolithic was linked with the local populations, and was widespread, and based, in the first place, but not always, on the richest sources of ores. Small deposits were useful as well, if they contained available oxide or carbonate copper minerals. Naturally, large deposits enabled further development of primary mining technology.

That is the reason one could not talk of a single main centre in southeast Europe, where extraction of copper ores had chronological priority. All those local centres were, probably, independent, and belonged to the same period, i. e. the Early and Late Eneolithic, and used broadly similar technology.

More exact knowledge of the mining technology has been derived from excavations carried out at Rudna Glava since 1968.

The prehistoric mine workings were here cut by a large open-cast iron mine, which includes underground works, too. On the north side of the open - cast iron mine were visible the shafts, partially destroyed, used at the time of the Vinča culture, i. e. during the Early Eneolithic of southeast Europe, with absolute dates of the first half or the middle of the fourth millenium B. C. (Plate 1 and 2).

It is possible now to say something more exactly about the zone of ore deposit, exploited during the Early Eneolithic. Although exact limits of the spread of the zone are not known, nevertheless we are dealing with a large deposit, principally of Malachite located on the south and south - west of the massif of Rudna Glava. This zone consists of numerous ore channels, often in the form of ore veins of various sizes, but also with broader cracks or fissures, spread longitudinally east - west. Inside this main ore deposit occur smaller zones, concentrated around fissures or the broad crack, where the ore bed had been more concentrated in the period of the formation of the deposit. The reliable chronological and spatial relations existed among those zones, as has been shown by the stratigraphy of the layers between the west platforms 3 - 5 (figure 1). It also implies an exploitation of long duration. However, all the zones have not yet been discovered, as current excavations at Rudna Glava are still of a small size. Quite enough data has however been obtained for the study of their function from the zone situated along the north and north east side of the open - cast iron mine and just partially preserved. The shafts in this zone were concentrated along a fissure, similar to the longitudinal line. (At the beginning of the description of the old mining workings, one should stress a terminological difficulty. In general mining terms such as shaft imply artificial structures made to reach the ore body. At Rudna Glava however, in the initial phase of mining, no structures of that kind existed, except access platforms. There are no traces of digging on the walls of the empty ore channels. In other words the first miners were able, according to the existing technology, to follow the natural channels of the veins and to extract its content. Therefore the term "shaft" is in this sense quite conditional and designates only the place where mining was carried out).

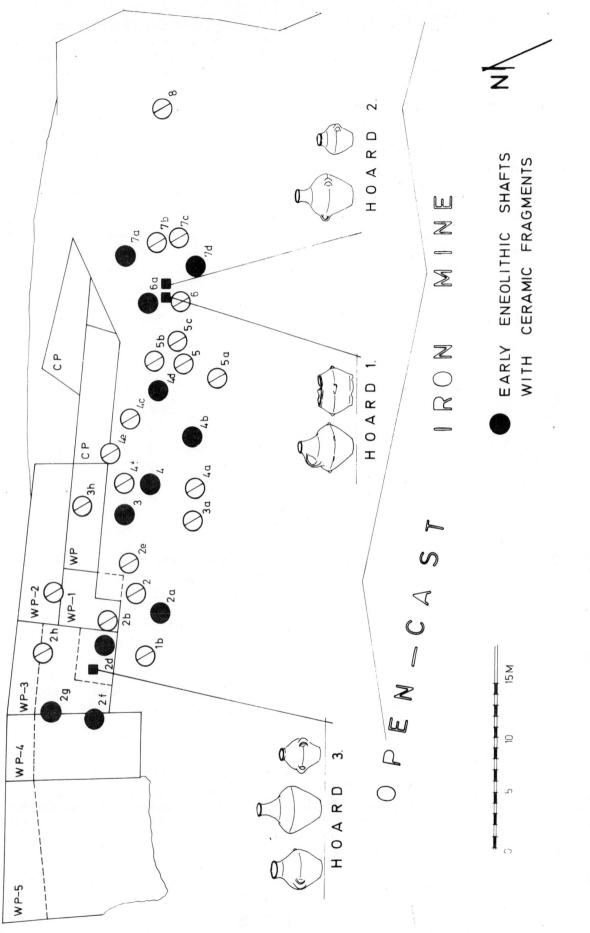

Figure 1 : Rudna Glava. Early Eneolithic shafts with ceramic fragments and the pottery hoards from access platforms.

It was confirmed by excavation, that all the veins along this fissure were very carefully exploited in the Early Eneolithic period. During this work the problem arose of disposing of the excavated material from the access platforms and the shafts, viz. moving the refuse material and then finding a suitable place for dumping it. The stratigraphy of the west platforms 3 - 5 has shown (figure 1), for example, that the mining activity here was discontinuous with exploitation temporarily transferred to other neighbouring veins.

During this interval a layer of a yellow, sterile earth accumulated. In the next, later level, represented mainly by dark earth, finds again appeared. This suggests that ore extraction was carried out immediately above these former used shafts. Later, some mining tools and ceramic fragments were brought down by erosion.

But excavated material was deposited from one shaft into another as well. The best proof for the gradual filling of empty shafts with material obtained from other areas were ceramic finds in the fill of the shafts (figure 1). For example, fragments of a pot found in the shaft 7d were found at different levels - this would not have been the case if the shaft had been empty before. It means that the pot was damaged and thrown away into a shaft partially filled up by blocks of limestone and remains of ores. The fragments fell through small cracks and hollows and finally came to different levels. From those fragments from shaft 7d a coarse ware vessel typical for the Vinča culture was later reconstructed.

Shafts 3 - 4 and 2f have not been fully excavated but they have shown the same disposition of the ceramic fragments and mining tools (figure 1, 2). It is understandable that the shafts could be filled step by step, or over longer intervals, or continuously, depending on the scale of mine workings carried out there. That filling consists of fragments of ore and limestone blocks, mixed with dark earth, sometimes similar to humus. It also implies another important characteristic of mining technology - the ores were crumbled or split near to the shafts and the useless remains were thrown back into the empty shafts. The ore in untouched veins is still very compact and no natural process could change this typical disposition of ores in the veins.

The same situation is repeated with the stratigraphy of stone mining mauls. These tools were usually put in places where the direction of the galleries changed, or bulged and in cracks suitable for tool stores. Those groups or single finds from some shafts should be understood as the tools left there at the end of the work (figure 2). Their position indicated the level of exploitation reached at this time by the Vinča miners. It also suggests seasonal mining work in the Early Eneolithic, so the different levels could be explained by temporary interruption of the mining. It seems that some groups of mauls were put on the level, that was the starting point for digging in the next mining season (spring - summer of the next year). In the case of continuous work it would not be necessary to leave tools on different levels, as the extraction would progress gradually to the bottom of the shaft.

The situation is different with fragments of tools, also found at different levels of the fill of the shafts (figure 2., Plate 3..) The explanation here is similar to the ceramic fragments - the pieces of mauls must have come with extracted material from the shafts, where the mauls were broken by work. Their recovery at different levels confirms again the gradual filling of an empty shaft. The tools were also found around the entrance in the shafts. They were usually just thrown away, but sometimes they lay in small groups. They were often damaged, so they could be regarded as reject tools from the nearest shafts. But they were left on the access platforms, with the possibility of secondary usage. Such examples of damaged mauls later repaired are otherwise known from Rudna Glava. One should also add that the mauls are simply pebbles, with a medial groove, the groove being the single feature of their treatment. Some pebbles without a groove were also found on the surface, around the access platforms transported here from the valley of the neighbouring river Šaška. Even though they were collected ready for preparation on the access platforms, (where such treatment was certainly done), these pebbles were not used. The mauls were just natural stones, and they represent the most

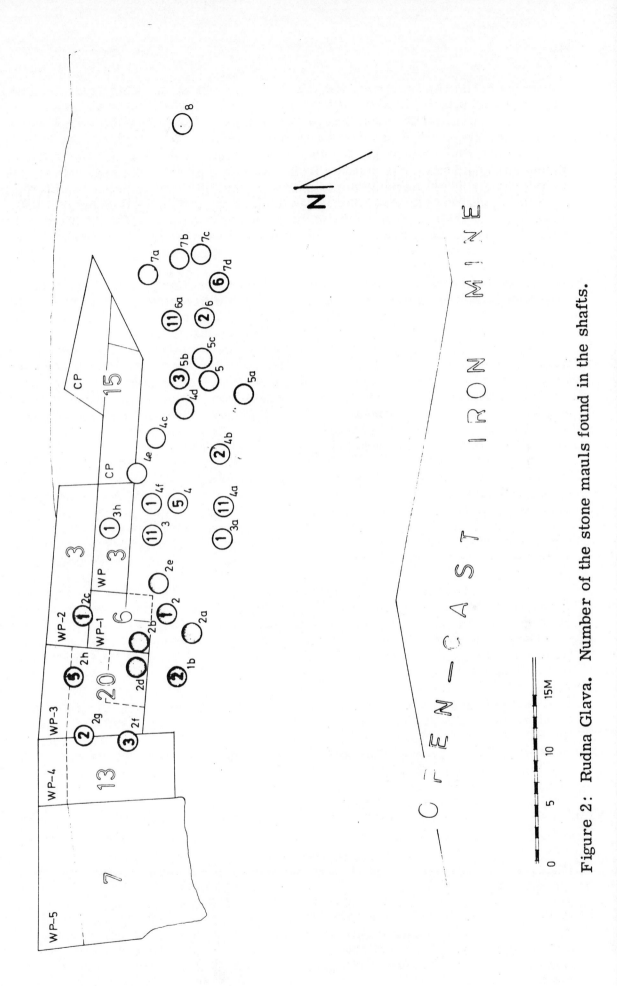

Figure 2: Rudna Glava. Number of the stone mauls found in the shafts.

numerous mining tools. Even nowadays it is possible to pick up, in a short time, quite enough pebbles of such kind in the river Šaška valley.

Pebble-mauls are nevertheless the results of a certain specialization, because there are general differences in their shapes. They can be divided between rectangular, oval, spherical, triangular and wedge shaped forms. To these typical forms one could add some smaller tools, without a groove, but still with visible wear traces. The difference in the shape and weight of those stone mauls implies also their different purposes during the work in the shafts. All of those mauls were used for the breaking and crumbling of ores, as attested by the visible damage on their working edges. Many mauls were often broken, because of very hard usage. A rope or vegetable roots, bound around the maul, gave the possibility of strong circular or vertical strokes in the very limited space of the shafts. The mauls are certainly the first specialized mining tools (Schmid, 1973), as they are never found in the settlements of the Early Eneolithic southeast Europe. They have appeared, as a rule, at places with the ore deposits (Jackson, 1971, and Rothenberg page 56 of this volume.)

Tools made of antler, wedges or some sort of scrapers with more tines, were also found in the shafts or at access platforms. Judging by the traces of wear, they served for the overturning and gathering the crushed ores. Together with possible wooden wedges, those of antler could be used for widening the cracks in ore blocks, which needed breaking up or crumbling. Similar tools are suggested at Siphnos (see Wagner et al, page 77 of this volume).

Wooden mining tools presumably were also numerous, but the wood in the shafts was generally very badly preserved. Some meagre traces of baulks were discovered near the access platform of the shaft 5b and 4d, along the north side of the longitudinal fissure. They belonged, maybe, to some sort of wooden construction, over which ore sacks were lifted with a rope.

The presence of such a wide range of tools, and the lack of any use of metal implements indicates without doubt that copper axes and hammers were not used in the oldest mining of copper. It is very well known that the copper industry, contemporary with the mine at Rudna Glava, already produced massive chisels and axes, as clearly shown by the well known hoards from Pločnik, south - east Serbia, (Jovanović, 1971). According to the data obtained from Rudna Glava they were tools for the domestic purposes, and also weapons.

The technique of the ore extraction by alternate heating and cooling is also represented at Rudna Glava. In some shafts scanty remains of fireplaces were discovered (shafts 4a, 2h). But the techniques used for ore separation is still not clear.

The form and appearance of copper carbonate minerals was, for example, very important at the beginning of the exploitation at Rudna Glava. Malachite and azurite were the results of disintegration of the chalcopyrite and were deposited, more or less, on the walls of channels of the veins, or along the numerous cracks and hollows in the ore body itself. But it would be very difficult to reconstruct an exact picture of the richest ore veins, as the shafts were cleaned very carefully both of copper minerals and magnetite. This precise work by the prehistoric miners is really noticeable and demands an explanation. (This has also been observed at Chinflon - Editor).

The separation of oxide copper minerals from the basic, magnetite ores had to be done directly at the mines, i.e. on the access platforms. It corresponds to similar activity in the choice of the best material in the flint mines in central and east Europe (Gurina, 1976). The copper ore at Rudna Glava was probably broken at the access platforms in small pieces and the very distinctive copper minerals were separated by mechanical treatment, viz. surface scratching. Even nowadays, a coating consisting of malachite or azurite, can be scratched from the walls of some shafts. It is the result of the process of oxidation of the crumbled ores in the fill.

This mechanical treatment could produce some concentrate, mixed with small pieces of magnetite. Further extraction could, maybe, be done by flotation.

Judging by the finds from the contemporary Eneolithic settlements in the Central Balkans, copper ores, consisting of oxide minerals, were further processed to a very fine powder (Jovanović, 1978). Such prepared ores could be very easily transported from Rudna Glava, as so far there has not been any evidence of smelting on or near the mine itself. The same situation is found at Ai Bunar (Chernych, 1978).

That the mining was a seasonal work is confirmed by the lack of any Vinča settlement not only in the vicinity of Rudna Glava, but also in a wider surrounding. The ceramic hoards from Rudna Glava reinforce this conclusion, as they consist of large vessels, not suitable for frequent transport. If any settlement existed near, it would not be necessary to hide the large vessels. Besides, the hoards, once hidden, remained untouched, as their owners, for some reason, did not return again. These large vessels, were used as containers for water, which was used to dowse the fines set against the rock and so split them.

Recent results obtained at Rudna Glava have shown the existence of certain relations concerning the ownership of some ore zones. It is also confirmed by the location of the ceramic hoards, which were distributed near the richest parts of the ore body. It is possible to think that organization of the mine corresponded with the system of production in the miners village. Or, in other words, the same property relations were represented both in the settlement and at the Rudna Glava mine.

Therefore the division of the rich ore zones is a question maybe, of a family or tribal property, rather than the common ownership of the whole village or several villages. Rather than the existence of separate miners settlements, the beginning of mining and metallurgy could be more closely linked with domestic activities. Thus the beginning of the specialization of the processing of metals could be sought in the activity of certain families or smaller tribes. It seems that the roots of the blacksmiths craft and other metal-smiths ought certainly to be looked for in Europe in the Early Eneolithic period. That was the time when important knowledge of mining and processing of metal had already been introduced.

All of the data, that have been discussed here, support the proposition of considerable production of copper during the Early Eneolithic. The number of the shafts investigated at present at Rudna Glava, their dimension and the depth, testify to the impressive quantity of the extracted ores. This quantity should be estimated in tons, even though one is ignorant of factors needed to calculate the precise amount.

Such considerable production of copper ores, in terms of absolute dating, marks the beginning of the technology of mining metal ores in South-East Europe. The exploitation was carried out by basic technical means, but quite satisfactorily adapted to the rich deposits. Therefore it is possible to explain a rapid increase in the quantity of metals in the Balkans and Danube Basin during the Eneolithic period, by the exploitation of rich, untouched and accessible deposits. A most important role was played by the simple, but effective mining technology.

This also implies that the stage of the exclusive usage of the native copper was very short, if it really existed as an independent phase in the history of the primary production of copper (Tylecote, 1976).

The exhaustion of the most suitable deposits with ore veins, rich in oxide and carbonate copper minerals marked a fall in the production of copper. The explanation for the rather low production of copper in the Late Eneolithic of south-east Europe, may well reflect these technological reasons, more than cultural ones. The transition to the mass production of the sulphide copper ores necessitated a shift in the contemporaneous centres of the metal industry further to the north - west to the Austrian Alps and the Central European ore mountains. But at the time of Rudna Glava, the primary exploitation of virgin deposits of oxide and carbonate copper ores flourished and caused a rapid development of the copper industry. The usage of copper in Prehistoric Europe was probably more directly dependent on the technological proficiency of mining, rather than on skill in processing the ore. It was the mining technology which replaced the exhausted deposits of copper carbonate minerals with those of sulphide ores, which gave to contemporary metallurgy a chance to discover a new, advanced method of mass production of metal.

REFERENCES

Chernych, N.E. 1978. Gornoe delo i metalurgija v drevnejshej Bolgarii. Sofija: Blgarska adademija na naukite.

Gurina, N.N. 1976. Drevnie kremnedobivajushcie shahti na teritorii SSSR. Leningrad: Nauka, leningradskoe otdelenije.

Jackson, J. 1971. Mining in Ireland. Technology Ireland October, 1-4.

Jovanović, B. 1971. Metallurgy of the Eneolithic Period in Yugoslavia. Beograd: Arheološki institut.

Jovanović, B. 1978. The Oldest Copper Metallurgy in the Balkans. Expedition, Vol. 21. No. 1, 9-17.

Schmid, E. 1973. Die Reviere urgeschichtlichen Silexbergbaus in Europe. Der Anschnitt Jahgg. 25/6, 25-28.

Pittioni, R. 1971. Prehistoric Copper Mining in Austria. Problems and Facts. Seventh Annual Report, Institute of Archaeology, University of London, 17-40.

Tylecote, R.F. 1976. A History of Metallurgy. London: The Metals Society.

Plate 1: Central part of the Early Eneolithic copper mine Rudna Glava, with the entrances, channels and bottoms of the partially destroyed shafts.

Plate 2:   Upper part of the destroyed shaft 6, Central part of the Early
            Eneolithic copper mine Rudna Glava.

Plate 3:   A pebble maul in situ in the upper part of the shaft 6a.

ANCIENT COPPER MINING AND SMELTING AT CHINFLON (HUELVA, SW SPAIN)

BENO ROTHENBERG AND A. BLANCO FREIJEIRO

1. Institute for Archaeo-metallurgical Studies, Institute of Archaeology, University of London, 31 Gordon Square, London WC1H OPY, England.
2. Universidad Complutense Madrid.

Abstract

In 1974, during an archaeo-metallurgical survey in the province of Huelva, remains of primitive mining and smelting were located at Chinflon. Excavations in 1976 and 1978 established three phases of mining, tentatively dated to Chalcolithic, Late Bronze and recent date. Remains of temporary habitation, seemingly in huts, were uncovered next to the mines, with pottery, stone implements and evidence for copper smelting and casting.

Slag nodules, collected on the surface and in the excavation, were found to be fayalite, non-tapped copper smelting slag. FeO flux was intentionally added. Small crucibles were used for melting-casting.

The significance of the Chinflon mines is enhanced by the proximity of a group of dolmens, apparently related to the ancient miners of Chinflon.

Keywords: COPPER, MINING, BRONZE AGE, SLAG, SMELTING, SPAIN, RADIOCARBON.

INTRODUCTION

In 1974 an archaeo-metallurgical research project was begun in the Province of Huelva (SW Spain), based on Rio Tinto and sponsored by Rio Tinto Zinc Ltd, London, and the two Spanish mining companies, Union Explosivos Rio Tinto and Rio Tinto Patino (today Rio Tinto Minera) operating the modern Rio Tinto Mines (Rothenberg, Blanco 1976). Its major objective is the study of ancient mining and metallurgy in South-West Spain, known through it's ancient slag heaps of many millions of tons as one of the major sites of ancient metal production. The Huelva Project is directed by Beno Rothenberg, on behalf of IAMS London, and Antonio Blanco Freijeiro of Madrid University. Up to date about 70 sites have been located and investigated in the province, including several excavations at key sites discovered in the preliminary surveys (Rothenberg, Blanco, 1976 and forthcoming).

HUELVA SURVEY 1974

Searching for mineral deposits suitable for early, prehistoric mining and smelting, we surveyed areas south and north of the major pyrite belt of Huelva, assuming that the huge gossan-capped ore bodies could not have been mined in prehistoric days and that the early metallurgists would have looked for minor and easily extractable oxidized ore deposits.

In the area of El Pozuelo, about 12 km south-west of Rio Tinto, a long chain of quartz outcrops runs across the hilly countryside. These hills, consisting of volcanic pyroclastic acid rocks, have recently been completely ploughed over and afforested with, still very small, eucalyptus trees, and the quartz outcrops stand out conspicuously as barren islands. The very first site visited in this area was the abandoned old mining site of Chinflon (HP 22), situated on top of a steep mountain, about 2km east of the village of El Pozuelo (figure 1 & 2). Here several quartz outcrops run along the upper part of the very steep hillside, with a mining shaft and a large tip heap as evidence of recent mining. In fact, several recent shafts were found penetrating the quartz outcrops, partly destroying ancient workings. In the outcrop veinlets of malachite can be seen, as well as traces of limonite and gossan. The mining dump contained bits of chalcopyrite, which occurs also in the recent galleries below and was indeed the main ore mine together with a little malachite, for a short period, at the beginning of this century (Pinedo Vara, 1963, 489).

In three of the quartz outcrops are ancient workings of a type also found at several similar sites in the area of El Pozuelo, but also further south, along the Rio Corumbel, southeast of Valverde del Camino, and at Cuchillares, north of Rio Tinto: Trench-like shafts (figures 2-4, Plates 1, 3), several metres long and about 80cm wide, penetrate into the white quartz body, with small rock-shelfs left across the trench as base for hauling devices. In some of these rock-bridges grooves were cut to hold wooden beams. The mining trenches are now mostly filled with soil and debris, but their shape indicate that the ancient miners just followed the oxidized ore into the depth of the outcrop.

There are only rough tool marks left on the walls of the trenches and it is obvious that only primitive blunt stone tools were used, many fragments of which were found both inside and outside the ancient mines. Indeed, the whole area of Chinflon was littered with innumerable grooved stone picks, made of hard, volcanic rock pebbles of various shape and size (Plate 1). The same types of implements were found at all similar trench-mining sites in the Huelva province, and pebbles for such use can be found in local river beds.

There were no slag heaps discernible at Chinflon, but our previous experience at early copper smelting sites (Rothenberg, 1972, 28; Rothenberg - Tylecote, 1978, 9) made us search for small nodules of crushed slag in the vicinity of the actual mining site (Plate 2). Indeed, a fair quantity of small slag nodules of a primitive type (see below) were found dispersed on top of the hill and on the slope above Mines 2 and 3.

Together with the slag we found a number of pottery fragments, mainly body-sherds but also a few rims and bases mainly of two different kinds: thick, roughly made bowl fragments and fine, well-burnished ware. At the time of our survey these sherds were identified as Chalcolithic and Late Bronze Age pottery, (Rothenberg-Blanco, 1976, 3).

COPPER MINES, SMELTING SLAG AND THE DOLMENS OF EL POZUELO

The old workings and the primitive smelting slag of Chinflon represent the earliest phase of mining and extractive metallurgy of copper found to date. The importance of this site is further enhanced by the proximity of a group of the famous dolmens of El Pozuelo (Leisner, 1959, 280-285 - N. 1-4), situated in the valley right below the mines. These are multiple elongated chambers, sometimes with transeptal arrangements, set in a circular mound, and their location in the desolate hills of this region, where no contemporary settlements nor any other sign of such early habitation has ever been found, had always been an enigma. According to the grave goods recovered, which also include a fragment of a copper awl (in N. 4) - (Leisner, 1956-59 Vol. I. 281-285, Plate 48) - the use of the dolmens began in the Neolithic phase and continued into the Early Copper Age, represented by the settlements and metallurgical centres of Los Millares and Vila Nova de Sao Pedro (Savory, 1968, 103, 152-162).

The proximity of the dolmens to the mining site and the archaeological and metallurgical finds at these sites, strongly suggested that the megalithic tomb

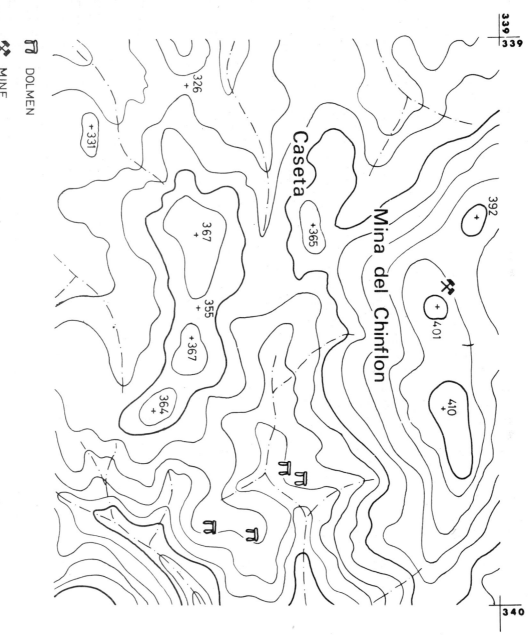

DOLMEN

MINE

Figure 1 : Chinflon and its dolmens (from Military Map of Spain, Sheet 960-1, Berracal, 1 : 10,000).

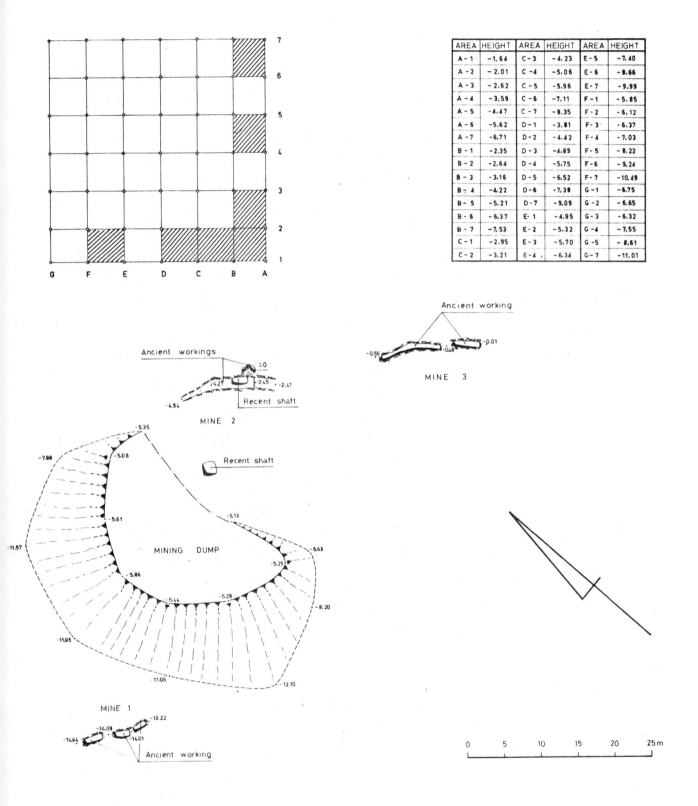

AREA	HEIGHT	AREA	HEIGHT	AREA	HEIGHT
A - 1	-1.64	C - 3	-4.23	E - 5	-7.40
A - 2	-2.01	C - 4	-5.06	E - 6	-8.66
A - 3	-2.62	C - 5	-5.96	E - 7	-9.99
A - 4	-3.59	C - 6	-7.11	F - 1	-5.85
A - 5	-4.47	C - 7	-8.35	F - 2	-6.12
A - 6	-5.62	D - 1	-3.81	F - 3	-6.37
A - 7	-6.71	D - 2	-4.42	F - 4	-7.03
B - 1	-2.35	D - 3	-4.89	F - 5	-8.22
B - 2	-2.64	D - 4	-5.75	F - 6	-9.24
B - 3	-3.16	D - 5	-6.52	F - 7	-10.49
B - 4	-4.22	D - 6	-7.30	G - 1	-6.75
B - 5	-5.21	D - 7	-9.09	G - 2	-6.65
B - 6	-6.37	E- 1	-4.95	G - 3	-6.32
B - 7	-7.53	E - 2	-5.32	G - 4	-7.55
C - 1	-2.95	E - 3	-5.70	G - 5	-8.61
C - 2	-3.21	E - 4	-6.34	G - 7	-11.01

Figure 2: Chinflon (HP 22). Plan of mining site and grid of excavations of the habitation area.

builders and users were in fact the earliest miners and metallurgists of Western Europe (Leisner, 1956-59, I, 13. I, /2. 280). However, a hypothesis of such magnitude could not be verified by a surface survey only and it was therefore decided to excavate at Chinflon.

EXCAVATIONS 1976-1978. (Figure 2)

1976

In 1976 a series of small trial trenches were dug on the slope above Mines 2-3, but with the exception of a small area near Mine 2, where the soil was held in place by a low rock outcrop crossing the slope, the depth of the soil on top of the slate rock surface was no more than 5-10cm, i. e. almost everthing was completely denuded. In trench X (part of Square B2 in grid of 1978), 4. 5 x 2. 5m, a hard brown surface was found at the depth of 25cm and here potsherds, slag nodules and fragments of mining hammers were found. Although most of these finds seemed not in situ, but washed in from higher up the slope, we assumed these to be habitation deposits, a view strengthened by the find of a small, carefully stone-lined post hole (No. 1), dug into the hard surface. Here we found again the same two kinds of pottery: handmade, rough bowl fragments and well-burnished Late Bronze Age pottery.

1978

We returned to Chinflon in 1978 to further investigate and excavate the mining shafts (B. Rothenberg with Phil. Andrews). The excavation of the adjacent habitation (see below) was directed by V. M. Hurtado Peres, D. M. Pellicer Catalan of Universidad Sevilla, and B. Rothenberg.

THE MINES

In the quartz outcrops three different mining technologies could be discerned, dated, tentatively, to the Chalcolithic - Early Copper Age (4th-3rd millennium B.C.), the Late Bronze Age (12th-8th century B. C.) and early 20th century A. D.

MINE 1 (figure 3, Plate 3)

This mine consists of three trench-like shafts of which shafts A and B were partly destroyed by recent workings, shaft C is an early trench-mine, now filled with boulders. Shaft A is of particular interest: At its west end, part of a round mining shaft was found in situ, carefully cut with metal chisels in short regular strokes, with a series of footholds in its sides, right down to as far as we could follow it without excavations (Plate 4). The originally round mining shaft of about 80cm diameter, was partly destroyed by a recent squarish shaft, which cut away half of the tubular shaft. Comparing this round, chisel-marked and footholded mining shaft with the mining technologies known from other mining sites (Wilson, 1977; Conrad-Rothenberg, forthcoming), and considering the periods represented at Chinflon by pottery finds, we propose for it a Late Bronze Age date.

It seems that only shafts B and C were originally chalcolithic trench-mines, for the extraction of malachite, whilst shaft A was a tubular Late Bronze Age mining shaft, probably leading down to the major chalcopyrite ore lode, which was again reworked in recent times. The small tip heap on the slope below Mine 1 seems of recent origin. Ancient grooved mining picks found right in the top layer of this dump are witnesses to the ancient origin of the mining shafts of Mine 1.

MINE 2 (figure 4)

This is a very large outcrop, cut into several sections by intensive trench-mining, now completely filled in. A recent mining shaft goes down for about 30m, with

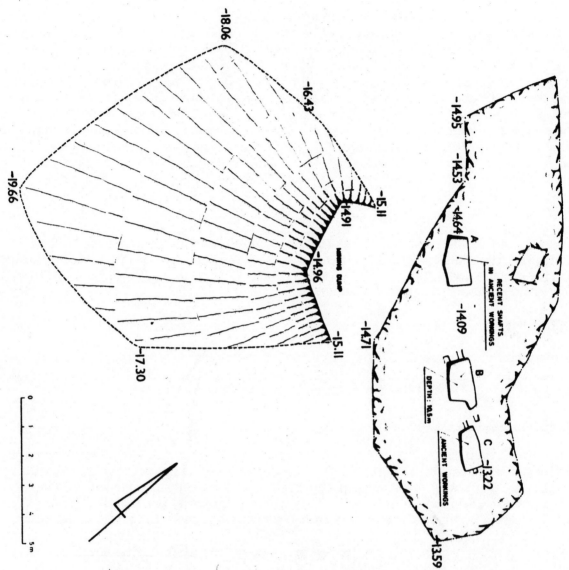

Figure 3: Chinflon (HP 22). Plan of Mine 1 and recent mining dump.

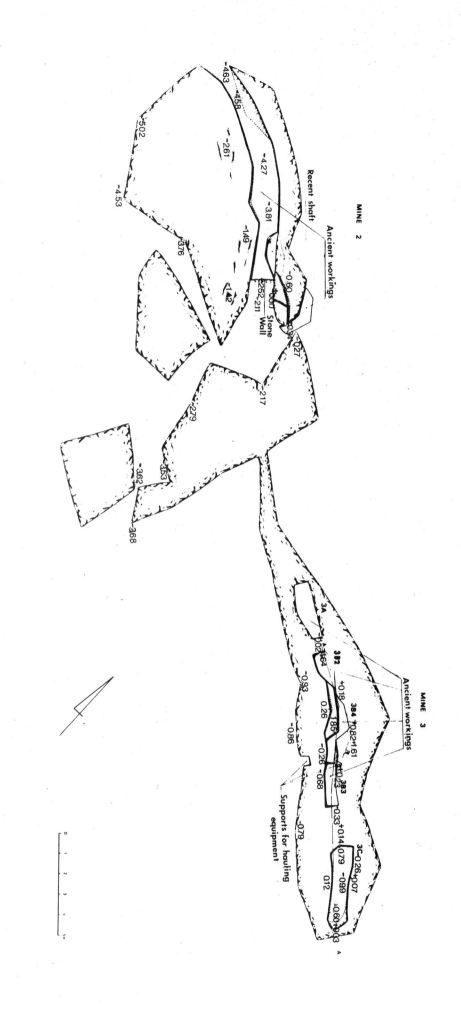

Figure 4: Chinflon (HP 22). Plan of Mines 2 and 3.

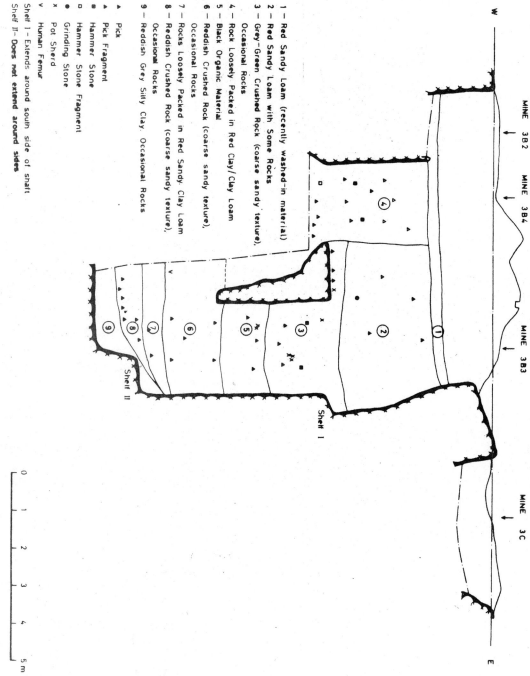

1 — Red Sandy Loam (recently washed-in material)
2 — Red Sandy Loam with Some Rocks
3 — Grey-Green Crushed Rock (coarse sandy texture), Occasional Rocks
4 — Rock Loosely Packed in Red Clay/Clay Loam
5 — Black Organic Material
6 — Reddish Crushed Rock (coarse sandy texture), Occasional Rocks
7 — Rocks Loosely Packed in Red Sandy Clay Loam
8 — Reddish Crushed Rock (coarse sandy texture), Occasional Rocks
9 — Reddish Grey Silty Clay, Occasional Rocks

▶ Pick
▲ Pick Fragment
■ Hammer Stone
□ Hammer Stone Fragment
● Grinding Stone
✕ Pot Sherd
∨ Human Femur

Shelf I — Extends around south side of shaft
Shelf II — Does not extend around sides

W

MINE 3B2   MINE 3B4   MINE 3B3   MINE 3C

E

0 1 2 3 4 5 m

Figure 5: Chinflon (HP 22).   Section A-A of Mine 3, excavated in 1978.

recent stone-walling protecting the mouth of the shaft from runoff rainwater.
(Pinedo Vara, 1963, 489).

THE EXCAVATION OF MINE 3 (figure 4).

This mine could only be partly excavated in 1978 - further work is planned for
1979 - but many details of its trench-mining technology have already been clarified.
There are three trench-like shafts in Mine 3 (Plate 5):   Shaft 3A is almost 3m
long, 60cm wide, and partly filled in.   A 70cm wide rockshelf separates shaft 3A
from shaft 3B, which is almost 8m long, at places extremely narrow (45cm) and
consists in fact of three separate mining shafts, separated by narrow rock shelfs
or ridges (3B2-4).   About 2m further to the east is shaft 3C, completely filled with
soil.   Square bosses were hammered out of the rock on both sides of these mining
trenches, apparently as bases for hauling devises.
    We excavated shaft 3B3 (Plate 6) and, partly, shaft 3B4.   These are in fact
adjacent shafts with a rockshelf separating them at a depth of 3.80m.   However,
this is not a real system of shafts and there are no proper galleries, but rather a
series of interconnected cavities hollowed out by extraction of the mineralized
rock.
    The stratification found in shafts 3B3-4 (figure 5) presents a far more com-
plicated picture as anticipated: Shaft 3B3 shows 9 layers of fill, but only the top
layer (1) and bottom layer (9) appear to be naturally deposited.   Layers (2)-(8)
consist of different materials - either very rocky (7), crushed rocks (3), (6), (8),
or organic debris (5) - which appear to have been intentionally backfilled into the
mine, perhaps from mining operations in shafts 3A or 3C.   The more or less even
distribution of stone picks and hammers in the layers (2)-(8) strengthen this interpretation of the layers in trench 3B3.
    Layers (2)-(9) must have originally continued also into shaft 3B4, but at some
time shaft 3B4 was at least partly reexcavated and the layers (2)-(3) disturbed or
removed - backfilled, at some later stage, with a mass (4) of mainly large rocks.
This fill, quite different from the layers in shaft 3B3, could however, be connected
with an ancient burial - not yet dated - as a human femur was found in layer (6)
underneath the rockshelf.
    A great number of complete and broken stone tools and some sherds were found
dispersed throughout the fill of shafts 3B3 and 3B4 and must therefore be considered
as not in situ.   However, some objects were found in situ on top of the rockshelf
between shaft 3B4 and shaft 3B3:   several complete grooved mining tools and part
of a handmade bowl of chalcolithic type, with a spout (figure 6, 2).   In 1979 shaft
3B4 was cleared and a small chalcolithic bowl was found (figure 6, 1).   Details of
the excavations in 1979 will be published in a forthcoming IAMS Mongraph.   In the
bottom of shaft 3B3 we found several wooden logs, perhaps rungs or fragments of
hauling beams.

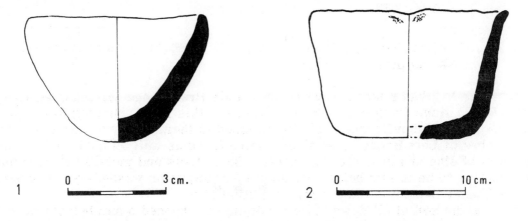

Figure 6:   Chinflon (HP 22).   Chalcolithic bowls found in Mine 3 B3-4.

THE HABITATION SITE

The following description is based on the forthcoming report of Pellicer Catalan and Hurtado Perez and further work done in 1979 by the authors. In 1978 the area around the trial trench of 1976 was excavated in order to uncover as much as possible of the habitation remains near Mine 2. Squares A-B-C 1, A2 and A0, were excavated and trial trenches dug in Squares E 1, A4 and A6. Clear habitation remains in situ were found only in a very limited area around Square B1 and it became evident that very little of the original habitation structures and installations had survived, in fact, the maximum depth of soil above the base rock was 30-40cm. Nevertheless, in Square B1 two distinct archaeological layers could be identified. Beneath the present surface (1) of soft, washed-in soil, overgrown by small shrubs and containing only very few finds, was a thin layer (2) of more solid texture but still consisting mainly of erosional fill, with some washed-in sherds, mining tool fragments and bits of slag. Beneath this a hard surface (3) was found with four stone-lined post-holes (figure 7).

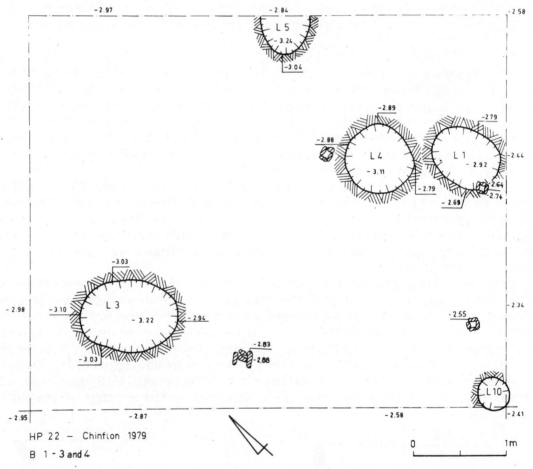

HP 22 — Chinflon 1979

B 1 - 3 and 4

Figure 7 :   Chinflon (HP 22). Plan of Square B1 level (3), with post holes and rock-cut pits.

Post-hole N. 2 had a large fragment of a Late Bronze Age jar inserted in its lining. Rough stone groupings were uncovered in this level, running in line across Square B1, but no significance could be attached to them. In level (3) a relatively large number of Late Bronze Age sherds were found as well as parts of broken picks, bits of slag and crucible fragments. Stone tools and metallurgical remains did not appear to be in situ but were either throwaways or washed-in. The area excavated was obviously a temporary habitation site.

Clearing the soil of (3) down to rock bottom (4) exposed a number of well cut round and oval pits (figure 8). These pits, 23-48cm deep, were found concealed by groups of small rocks (Plate 7) and contained a number of sherds and fragments of mining picks. In pit L. 4 a quantity of red clay was found sticking to the sides and bottom and it may well have been a storage pit for refractory material.

There were no proper habitation remains at level (4) and the rock-cut pits appear to be evidence for an open air working and storage space such as are common next to metal producing installations at sites of extractive metallurgy. However there was sufficient stratigraphic evidence to demonstrate the existence of two distinct archaeological layers, the upper one a habitation surface with four post-holes and the lower one represented by the group of rock-cut pits: Clearing the soil right next to post-holes N. 1 and N. 2 pits were uncovered right underneath the post-holes (Plate 8), containing, besides pottery, several complete mining picks which appeared to have been intentionally put there. The very location of the post-holes make it seem most unlikely that the makers of the post-hole knew anything about the earlier pits underneath.

The problem of dating the bottom layer of the excavation was enhanced by two facts: first, because of the heavy erosion the bottom layer could not be well identified and separated over most of the area excavated, especially as it seems to have been an open working and storage space without any architecture, and second, rough handmade pottery of Chalcolithic-Copper Age tradition still occurs in excavations of south west Spain in Late Bronze Age context. Furthermore, the pottery found in the pits of level (4) in Square B1 as well as on the rock surface of some of the other squares, consisted mainly of rough handmade bowl fragments with few characteristic shapes which could not be accurately dated by pottery typology. However, the ware of the sherds found in the rock-cut pits of level (4) is rather different, in matrix and temper, from the Late Bronze Age rough ware of Huelva and is very similar to Chalcolithic-Copper Age ware. (A large quantity of LBA pottery ware recently found by the Huelva Archaeo-metallurgical Project's excavations in the bottom layer of the Rio Tinto slag heap and it's comparison with the pottery from Chinflon proved the above without any reservation. A petrographic study of the Chinflon and Rio Tinto pottery will be published in the forthcoming excavation report on Chinflon).

MINING TOOLS AND STONE AND FLINT OBJECTS

The dating of the grooved mauls and picks found at Chinflon can unfortunately not be based on secure stratified data from the excavations. Yet, we nevertheless propose an early, Chalcolithic-Early Copper Age date for these mining picks. Though very many stone tools were found discarded and in fragments all over the site and also in the fill of Mine 3 and in all levels of the excavations in the habitation area, the only complete tools came from the bottom and the rock shelf of Mine 3 and from the pits in the lowest level of Square A1 and B1. The stone tool fragments found in the excavation appeared to be mainly broken throwaways, re-used sometimes, according to the excavators, as stray stones, but all complete and serviceable tools were only found in connection with the earliest technological phase of the Chinflon mine and it's copper metallurgy for which we propose a Chalcolithic-Early Copper Age date.

There are numerous comparisons for the use of large grooved pebbles of volcanic hard rock as mining tools in mines of the Chalcolithic-Early Copper Age cultural horizon, for example at Rudna Glava in Jugoslavia (Jovanović, 1971; 1974; 1978; forthcoming), Aibunar in Bulgaria (Černych, 1978), at Mount Gabriel in Ireland (Jackson, 1971) in the Veshnoneh area in West-Central Iran (Holzer et al, 1971), Timna in Israel (Rothenberg, 1972; 1978), and many other sites across Central Europe, Asia Minor and the Far East. Furthermore, we consider it of decisive importance for the early dating of the mining tools of Chinflon, that no grooved mining mauls, picks or hammers were ever found in any of the Late Bronze Age mining sites surveyed and excavated in the Huelva province, including Rio Tinto, where extensive Late Bronze Age remains were found right on the site of ancient mining (Rothenberg, Blanco, forthcoming). There can be little doubt, that the Late Bronze mining technology would require the use of metal mining tools -

bronze picks and chisels - and that the use of primitive stone mauls, picks and hammers as major mining tools belongs essentially to a very early phase of pre-historic technology.

One complete, polished stone axe and the fragment of a second, of a type extremely common in the dolmens, were found near the mines of Chinflon, but this type is to be found, according to M. Pellicer and V. Hurtado, from the Neolithic to the Late Bronze Age and does not, by itself assist in dating the mining site.

A number of flint artifacts and some waste of flint working were found in the excavation of the working and habitation areas, including one core and one arrow-head of a type well-known from the dolmens of Pozuelo. As flint is not native in Chinflon and it's surroundings, it must have been brought to the site by the early miners. Because flint implements were no longer used in the Late Bronze Age, they must have belonged to the Chalcolithic-Early Copper Age metallurgists of Chinflon.

METALLURGY (Table 1)

The slag nodules found at Chinflon (Plate 2) represent a very primitive stage of smelting technology, well-known from Chalcolithic smelting sites in the Arabah and Sinai (Rothenberg 1978,9). It is a very inhomogeneous, porous and rather viscous fayalite slag, though the smelting charge must have contained intentionally added iron-oxides to act as a flux, besides malachite. We know from other early copper smelting sites, that fluxing was already a well-established practise in Late Chalcolithic and Early Bronze times (Rothenberg, 1979; Lupu, 1970). The slag was not tapped out of the furnace but mechanically removed from the smelting furnace in big lumps and crushed to small nodules in order to extract the entrapped copper prills. No furnace lining nor tuyeres were found at Chinflon and the smelting furnace must have been of the hole-in-the-ground type, operated by bellows of animal skin.

The extractive process used to produce the slag found at Chinflon reflects in fact the earliest, primitive copper smelting technology known to date (Rothenberg-Tylecote, 1978; Lupu, 1970, 21-23) and belongs to the Chalcolithic phase of ex-tractive metallurgy. No slag nodules of this primitive type were found at any of the Late Bronze Age smelting sites in Huelva (Rothenberg, Blanco, 1976, 6-7), or, for that matter, in any of the Late Bronze Age sites investigated in the ancient world (Tylecote, 1976, 29-39). All copper smelting sites at Late Bronze Age technological level, i.e. industrial copper smelting, show proper tapped slag in considerable quantities and are always accompanied by furnace lining and tuyeres. This has now (1979) been confirmed by the excavations at Corte Lago in Rio Tinto, where extensive Late Bronze Age workings show a very high development of smelting technology, including proper furnace lining and tapping facilities.

None of the slag at Chinflon was found in situ, i.e. the slag nodules were washed in from a nearby smelting installation, outside the habitation and storage area excavated. The very small quantity of primitive slag nodules present at Chinflon, as compared with the relatively large deposits and heaps of tapped slag at the Late Bronze Age sites of Huelva, also suggests a date close to the very beginning of copper smelting. The scale of copper production at Chinflon was in the range of tens of kilograms - a valuable production at a time copper was scarce and almost considered a precious metal - as compared with tons of metal produced in the Late Bronze Age copper industries. No slag of possible Late Bronze Age type was found at Chinflon and we must assume, that the Late Bronze Age miners of Chinflon either transported their ores to a distant smelting site, or never reached an economic stage of ore extraction and shaft A in Mine 1 represents an exploratory operation only. Indeed many purely exploratory shafts of various ages can be found all over the mineral belt of Huelva, with no evidence of actual ore extraction or metal production associated with them.

Table 1: Copper ores and smelting slags.

ORES   (EDXRF – %)   (H.G. Bachmann)

Sample no	SiO$_2$	Cu	FeO	Pb	Zn	MnO	Al$_4$O$_3$	S	CaO	P$_2$O$_5$	K$_2$O	TiO	
310	59	40.3	35.7	—	0.4	0.3	—	4.2 (?)	0.3	1.2	0.2	0.2	Recent mine (chalcopyrite)
315	41.9	26.9	13.6	—	0.2	0.3	0.2	2.6	0.3	1.0	0.6	0.1	Ancient mine (malachite)
314	2.8	1.4	80.3	0.2	0.1	0.5	1.4	0.4	0.4	2.1	0.3	0.4	Gossan (hematite)

SLAGS   (A.A.–%)   (Geomet Service)

Sample no	SiO$_2$	Cu	Fe	Pb	Zn	Mn	Al	S	Ca	As	Ag.	Ni	Co	Bi	Na	Sb
233	22.80	2.64	44.0	0.03	0.15	0.11	1.99					0.005	0.06	0.020	<0.05	<0.010
234	18.10	1.19	50.2	0.05	0.79	0.12	1.90					0.005	0.11	<0.015	<0.05	<0.010
235	14.60	5.90	51.0	1.04	0.80	0.06	0.60					0.010	0.07	0.020	<0.05	<0.015
236	20.30	1.18	48.0	0.04	0.12	0.24	1.62					<0.001	0.29	<0.015	<0.05	<0.010
238	20.50	2.72	49.8	0.03	0.26	0.11	1.50					<0.001	0.07	0.020	0.18	<0.010
239	29.90	3.76	24.1	0.05	0.04	0.44	1.50					0.010	0.03	<0.015	<0.05	<0.010
240	32.60	1.95	24.8	0.06	0.41	0.10	1.68					0.010	0.04	<0.015	<0.05	<0.010
241	18.80	1.78	49.1	0.15	0.89	0.15	1.07					0.005	0.18	<0.015	<0.05	<0.010
242A	37.30	1.25	34.8	0.10	0.16	0.10	0.57					0.001	0.08	<0.015	<0.05	<0.010
243A	15.60	0.91	53.1	0.05	0.08	0.09	1.24					0.010	0.28	<0.015	<0.05	<0.010
319A	24.60	2.21	48.0	0.06	0.15	0.09	1.85	0.308	0.12	500	164	0.006	0.07	0.014	0.028	0.002
320	20.18	1.54	50.8	0.10	0.25	0.08	0.23	0.456	0.12	200	11	0.018	0.07	0.012	0.032	0.004
321A	7.79	2.07	70.6	0.06	0.04	0.01	1.04	0.104	0.02	350	11	0.005	0.03	0.034	0.005	0.005

SOME C14 AND THERMOLUMINISCENCE DATES

Charcoal samples for radiocarbon dating were collected from levels 1-4 of the habitation and from Mine 3. At present four dates are available - 5570 year half life for C14 (R. Burleigh, British Museum Research Laboratory, 8.3.1979 1, for calibration of the dates see Clark 1975):

Sample HP 270 Square A1 level 2   BM 1528 2650 $\pm$ 60 bp (700bc) - 880 B.C.

Sample HP 267 Square B1 level 3   BM 1529 3320 $\pm$ 130 bp (1370 bc) - 1680 B.C.

Sample HP 435 Square B (Locus 4, level 4) BM 1600 2890 $\pm$ 506 bp (940bc) - 1160 B.C.

Sample HP 436 Mine 3 Layer 9. BM 1599 2830 $\pm$ 50 bp (880bc) - 1070 B.C.

Whilst samples HP 270, 435 and 22 suggest the Late Bronze Age, sample HP 267 (BM 1529) labeled B1 level 3, consisted of a mixture of tiny bits of charcoal collected over a wide area in the lower levels of Square B1 and does not really represent the lowest deposits at Chinflon. Yet, it can serve as a reliable indication that the habitation at Chinflon began long before the Late Bronze Age.

A few sherds from level 2 in Square B1 were submitted for thermoluminescence dating (Joan Huxtable - M. Aitken, Oxford Research Laboratory for Archaeology, 30. 11. 1978):

OXTL 200e 3   530 B. C. $\pm$ 175

This result conforms well to the C14 date obtained for the same level.

Several sherds from level 3 of Square B1, obtained in 1978, gave the following date (the mean of four sherds):

OXTL   200e 3 (II)   2050 B. C. $\pm$ 300

Further sherds, from pits L. 1 and L. 4 are being processed at present and should further assist in our dating of the earliest phase of Chinflon.

M. J. Aitken (24. 9. 1974) dated a sherd from a dolmen at Los Gabrieles (near Valverde del Camino), excavated in 1974 (Rothenberg-Blanco, forthcoming) which was of the same type of thick, handmade pottery collected at Chinflon during the survey. Aitken writes (OXTL 199c): 'It could be as old as 3000 B. C. , but it's not possible to be more definite than this without soil samples and details of burial'. According to subsequent verbal communication (Nov. 74) this sherd could well be of the 4th millennium B. C.

CONCLUSIONS

The earliest phase of copper mining and smelting activities at Chinflon present the picture of a primitive copper industry belonging to the cultural horizon of the Chalcolithic-Early Copper Age. This phase appears to be connected with the megalithic burial site, the dolmens of El Pozuelo situated immediately adjacent to the mining site.

A later phase of mining was dated to the Late Bronze Age, but no copper smelting took place during this period at Chinflon.

Right next to the copper mines a small Chalcolithic-Early Copper Age working and storage site was excavated. No metallurgical activities took place within the excavated area but small slag nodules and crucible fragments, apparently washed-in from higher up the slope, indicate copper smelting and casting activities nearby.

The abandoned Chalcolithic-Early Copper Age working site was again occupied during the Late Bronze Age by temporary habitation in huts or tents, probably by Late Bronze Age miners, but very few remains of this occupation remained in situ.

On the evidence available at this stage of our investigations we suggest for Chinflon a date in the 3rd millennium B. C. for the earliest phase of mining and smelting and the 12-8 cent. B. C. for the second phase of mining. The last mining activities at Chinflon took place at the beginning of the present century.

REFERENCES

Černych, E. N.   1978.   Aibunar, a Balkan copper mine of the fourth
                                     millennium B. C. , Proceedings of the Prehistoric
                                     Society, 44, 203-217, plate 15-20.

Clark, R. M.   1975.   A calibration curve for Radiocarbon dates
                                     Antiquity 49   151-266.

Conrad, H.G. and Rothenberg, B. (forthcoming)		Antiker Kupferbergbau und-verhuettung im Timna-Tal, Israel, Anschnitt, (Sonderheft), Bochum
Holzer, F.H., Momenzadeh, M. and Groop, G.	1971.	Ancient Copper Mines in the Veshnoveh Area, Kuhestan-E-Qom, West-Central Iran, Archaeologia Austriaca, 49, 1-12.
Jackson, S.J.	1968.	Bronze Age Copper Mines on Mount Gabriel, West County Cork, Ireland, Archaeologia Austriaca, 43, 92-114.
Jovanović, B.	1971.	Metallurgy of the Eneolithic Period in Jugoslavia, Belgrade.
	1974.	Technologie minière de l'Eneolithique Ancient centre-balkanique, Starinar, 23, 1972, 1-14, plate 1-6.
	1978.	The Oldest Copper Metallurgy in the Balkans, Expedition, 9-17.
(forthcoming)		Rudna Glava, the Earliest Copper Mining Site of Eastern Europe, Archaeo-Metallurgy, IAMS Monographs, London.
Leisner, G and V.	1956- 1959.	Die Megalithgraeber der Iberischen Halbinsel, Madrider Forschungen, 1, Berlin.
Pinedo Vara, I.	1963.	Piritas de Huelva, Summa, Madrid.
Rothenberg, B.	1972.	Timna, Thames and Hudson, London.
	1978.	Excavations at Timna Site 39, Archaeo-Metallurgy, IAMS Monographs 1, 1-15.
	1979.	Sinai, Kuemmerly & Frey, Bern.
Rothenberg, B. and Blanco Freijeiro, A.	1976.	The Huelva Archaeo-Metallurgical Project, IAMS, London.
(forthcoming)		Ancient Mining and Metallurgy in the Province of Huelva, SW Spain, Archaeo-Metallurgy, IAMS Monographs.
Rothenberg, B., Tylecote, R.F. and Boydell, P.S.	1978.	Chalcolithic Copper Smelting, Archaeo-Metallurgy, IAMS Monographs 1, London.
Savory, H.N.	1968.	Spain and Portugal, Thames & Hudson, London.
Tylecote, R.F.	1976.	A History of Metallurgy, The Metal Society, London.
Wilson, A.	1977.	Timna, Mining Magazine, 262-271.

Plate 1: Chinflon (HP 22). Grooved mining pick.

Plate 2: Chinflon (HP 22). Slag nodules of Chalcolithic type.

Plate 3: Chinflon (HP 22). Mining trench of Mine 1, shafts B-C. A rockshelf separates the individual shafts.

Plate 4: Chinflon (HP 22). Late Bronze Age tubular shaft with footholds and
chisel marks.

Plate 5: Chinflon (HP 22). Mine 3A - a typical trench mine.

Plate 6: Chinflon (HP 22). Trench mine 3B3 during excavation.

Plate 7: Chinflon (HP 22). Habitation site, Square B1, with rock-filled pits in bottom layer (4).

Plate 8:   Chinflon (HP 22).   Area B1 - Stone-lined post hole of level (3), in upper left corner, situated high above a pit of level (4).   The pit contained mining picks.

EARLY BRONZE AGE LEAD-SILVER MINING AND METALLURGY IN THE AEGEAN: THE ANCIENT WORKINGS ON SIPHNOS

G. A. WAGNER, W. GENTNER, [1] H. GROPENGIESSER [2]
AND N. H. GALE. [3]

1. Max-Planck-Institut für Kernphysik, Saupfercheckweg, Heidelberg.
2. Archäologisches Institut der Universitat, Marstallhof, Heidelberg.
3. Department of Geology and Mineralogy, Parks Road, Oxford.

Abstract

On Siphnos island lead-silver mineralizations, mainly as complex antimony sulphosalts, occur within extensive iron ores. The lead-silver ores from various sites on Siphnos have very similar mineralogical, chemical and lead isotopic compositions indicating their genetic relationship. Ancient workings in these lead-silver ores have been observed at Ayos Sostis, Ayos Silvestros, Vorini, Kapsalos and Xeroxylon. The occurrence of litharge at Ayos Sostis, Kapsalos and Plati Yialos proves that cupellation was practised. Radiocarbon, thermoluminescence and archaeological studies date the workings at Ayos Sostis to the Early Cycladic and the Archaic periods. Lead isotope compositions of some Early Cycladic lead artefacts found on other islands indicates that the metal came from Siphnos. This evidence implies that the Aegean region must have played a much more active role in lead-silver metallurgy during the Early Bronze Age than was previously known.

Keywords: EARLY BRONZE AGE, EARLY CYCLADIC, ARCHAIC, SIPHNOS, AEGEAN, ORES, LEAD, SILVER, MINING, CUPELLATION, X-RAY FLUORESCENCE, NEUTRON-ACTIVATION, LEAD-ISOTOPES, RADIOCARDON, THERMOLUMINESCENCE.

INTRODUCTION

The Cycladic island of Siphnos is of considerable interest in early metallurgy. Several ancient writers mention this island in connection with gold and silver. Herodotus (III, 57) was the first to report, in the 5th century B.C., on the rich gold and silver mines of Siphnos. With the tithe of their revenue the Siphnians built their famous treasury at the shrine of Apollo in Delphi. In the 2nd century A.D., Pausanias (X, 11) wrote that Siphnos once had gold mines which were flooded by the sea and obliterated when the Siphnians gave up their tribute to Delphi. The 10th century Byzantine lexicographer Suda tells a similar story about the mines but mentions only silver. Gold and silver as the source of the past wealth of the Siphnians was mentioned by Eustathios in the 12th century, but there is no hint, in the classical authors, that the mines were exploited prior to the 7th century B.C., Gale (1979b) has reviewed the observations about ancient mining on Siphnos made by 18th and 19th century visitors to the island. By the end of the 19th century most of the ancient Siphnian mines had been re-opened, but instead of gold and silver the modern miners worked iron ores only. The fact that neither gold nor silver

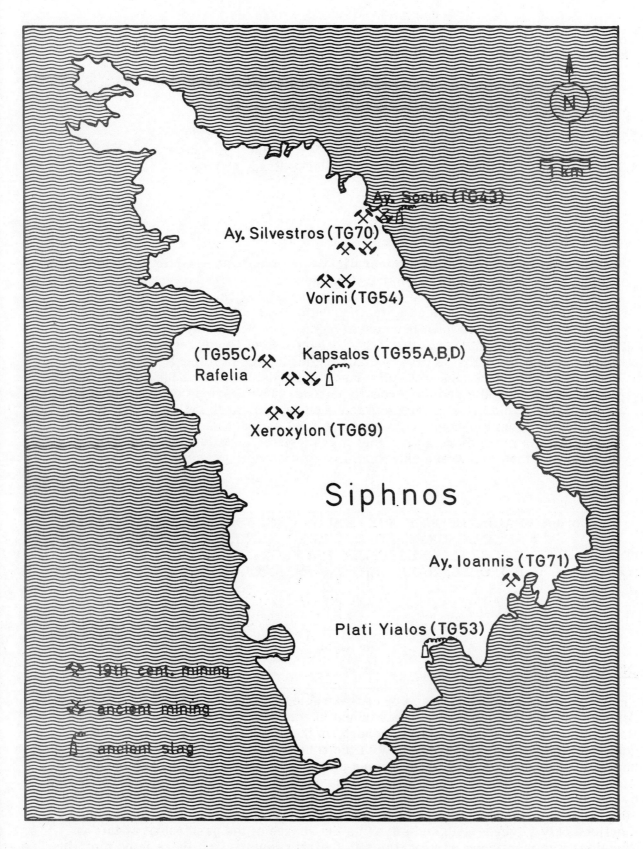

Figure 1: Ancient and modern workings on Siphnos.

Table 1: Brief description of analyzed ore samples from Siphnos.

TG 43-10 Ayos Sostis, ancient mine 2[+], cerussite fragment from under-ground waste-ore ("Versatz")

TG 43-11 Ayos Sostis, out-cropping, "ore-vents", yellow mineralization included within iron-manganese ores

TG 43-13 Ayos Sostis, collapsed ancient gallery, yellow mineralization from wall of gallery with traces of ancient mining

TG 43-15 Ayos Sostis, ancient mine 2, green ore fragment from "Versatz"

TG 43-16 Ayos Sostis, ancient mine 2, grey-yellow-red ore fragment from "Versatz"

TG 43-19 Ayos Sostis, Out-cropping "ore-vents", green-blue mineralization included within iron-manganese ores

TG 43-20 Ayos Sostis, ancient surface dump, green ore fragment from above-ground waste-ore

TG 43-22 Ayos Sostis, ancient surface dump, yellow ore fragment from above-ground waste-ore

TG 43-36.1 Ayos Sostis, ancient mine 4, yellow mineralization (from wall of ancient gallery)

-36.2 Ayos Sostis, ancient mine 4, brown mineralization (from wall of ancient gallery)

-36.3 Ayos Sostis, ancient mine 4, blue-green mineralization from roof of ancient gallery

TG 43-52.2 Ayos Sostis, ancient surface dump, yellow ore fragment from above-ground waste-ore

TG 43-52.3 Ayos Sostis, ancient mine 2, yellow-green ore fragment from "Versatz"

TG 43-56.2 Ayos Sostis, ancient surface dump, yellow-green ore fragment from above-ground waste-ore

TG 43-57.2 Ayos Sostis, ancient mine 2, blue ore veinlet from wall of gallery

TG 43-59.3 Ayos Sostis, ancient surface dump, yellow ore fragment from above-ground waste-ore

TG 43-62.2 Ayos Sostis, outcrop, yellow mineralization included in iron ores

TG 54-3.2 Vorini, ancient gallery, yellow ore fragment from "Versatz"

TG 54-4.2 Vorini, same gallery as TG 54-3.2, yellow mineralization included in iron-ores from wall of gallery

TG 55A-5.2 Kapsalos-Frase, ancient gallery, green-yellow ore fragment from "Versatz"

TG 55A-6.2 Kapsalos-Frase, modern pit, yellow mineralization included within iron ores from wall of modern mining pit

TG 55A-7.3 Kapsalos-Frase, ancient gallery, yellow ore fragment from "Versatz"

TG 55A-7.4 Kapsalos-Frase, same gallery as TG 55A-7.3, yellow-green ore fragment from "Versatz"

TG 55D Kapsalos-Tsingoura, modern mine, yellow mineralization included within iron ore from wall of modern mining pit

TG 69-1.2 Xeroxylon, modern surface dump, yellow-green ore fragments

TG 70-1.2 -1.3 Ayos Silvestros, modern pit, very heavy, grey-green-yellow mineralization included within iron ore from wall of modern mining pit

[+]for localities within the Ayos Sostis region see Figure 3 by Wagner and Weisgerber (1979)

Table 2: X-ray fluorescence analyses (in %)+) of various ores from Ayos Sostis.

	SiO_2	P_2O_5	K_2O	CaO	TiO_2	MnO	FeO	Cu	Zn	As	Pb	Sb
TG 43-10	7.2	0.7	0	4.7	0	0.2	0.7	0.3	0.2	0	83.9	0
TG 43-11	5.9	3.9	0	1.8	0	0.5	15.4	0.3	0.3	6.4	58.9	0
TG 43-13	3.2	0	0	2.4	0	0.4	4.2	0.6	0	0	38.3	4.0
TG 43-15	n.d++)	n.d.	n.d.	10.2	0	0.9	3.5	37.5	2.2	0	2.3	1.8
TG 43-16	0	0	0	2.8	0.1	1.2	38.5	0.2	30.9	0.1	1.3	1.4
TG 43-19	6.9	1.1	0	4.7	0	0.9	7.7	0.5	0.4	0.8	67.6	0.8
TG 43-20	1.7	0	0	32.2	0.1	0.2	3.9	14.4	15.9	0.3	1.8	0.2
TG 43-22	3.0	0.3	0.2	27.5	0.1	0.2	1.1	0.3	0.7	1.3	25.6	23.6
TG 43-36.1	3.6	0	0	8.2	0	3.3	4.1	13.9	1.5	0.1	34.5	20.4
TG 43-36.2	0	0	0	12.5	0.1	4.8	41.0	1.0	0.5	0	5.4	1.7
TG 43-36.3	2.1	0.3	0.5	30.0	0.1	0.2	0.2	5.3	1.1	6.8	48.1	1.5

+) XRF analyses were performed by Dr. Bachmann, Degussa Werk Wolfgang

++) n.d. = not determined

were any more observed cast some geological doubts on the ancient reports of rich gold and silver mines on Siphnos, doubts which the small scale of the Siphnian coinage did little to allay.

In order to solve these uncertainties about the ancient workings on Siphnos several field excursions to Siphnos were undertaken between 1975 and 1978. The aim of our field and laboratory studies was to localize as many as possible of the ancient mines, to establish which metals were produced from them and where they were used, to date the workings and to learn about their significance in antiquity.

Previously the Siphnian ancient gold and silver mines were associated mainly with the Ayos Sostis site (Figure 1). There, near the sea shore, numerous ancient galleries are visible. The galleries are flooded in their lower parts. This supports Pausanias' statement. In a recent report (Wagner and Weisgerber, 1979) this site was identified as an ancient mining site of lead and silver. Ancient workings on Siphnos were known also from Kapsalos, from the slope of Oros Profitis Elias - which probably means the site of Xeroxylon - and from Vorini (Gale, 1979b). In addition to these sites traces of ancient workings were observed also on the Ayos Silvestros mountain (Figure 1). Actually, nearly all the modern iron mines on Siphnos were developed from ancient galleries. The modern miners simply blasted open the ancient workings, largely destroying them. Therefore, remains of ancient workings may be found around the modern iron pits.

CHEMICAL ANALYSES

All the sites mentioned lie along an iron-manganese mineralization belt which crosses the island in a NE-SW direction. The iron and manganese oxides and carbonates occur in brecciated zones of the Tertiary marble. Occasionally this mineralization contains small yellowish clusters up to about 10cm size which consist of brittle, heavy, fine grained material. These inclusions are lead-silver ores. It is obvious from the arrangement of the ancient galleries that the ancient miner was not interested in the common iron ore but was mainly looking for these inclusions.

Microscopic studies on polished sections of such inclusions revealed the rather complex nature of these ores. The small grain-size makes the microscopic identification of the minerals very difficult. X-Ray diffraction studies were carried out on powdered ore fractions. The following minerals were identified: the secondary lead minerals cerussite and anglesite, the secondary copper minerals malachite and azurite, the secondary zinc mineral smithsonite, the antimony sulphosalts tetrahedrite, stephanite and robinsonite, the lead-antimony oxide bindheimite, and a little galena. This highly oxidized mineralization essentially was observed at all sites. These ore deposits must form part of the leached iron hat of primary sulphide zones hidden at depth. Previous work on the mineralogy of the Siphnian ores appears in Klein, (1979).

Numerous ore samples which are briefly described in table 1 were analysed chemically (see Müller and Gentner, (1979), for previous analyses). Gold, silver and antimony contents were determined by neutron activation. The antimony content of the ores is rather high and ranges up to 24.9% (in sample TG 54-4.2). Because the antimony disturbs the analysis of the other elements it has to be radiochemically separated. The gold and silver contents of the ores are plotted in figure 2. In addition to neutron activation some of the ore samples from Ayos Sostis were also analysed with X-ray fluorescence by Dr. Bachmann (Degussa Werk Wolfgang).

The results of the XRF analyses are given in table 2. The high lead and antimony contents indicate that the yellow ore inclusions consist essentially of complex lead-antimony-sulphosalts with some copper and silver. The occasionally high iron contents are caused by contamination from the surrounding iron ore. The most interesting result of these chemical analyses is the silver content of about 0.1% to 0.5% in ores from all ancient mining sites on Siphnos. This compares well

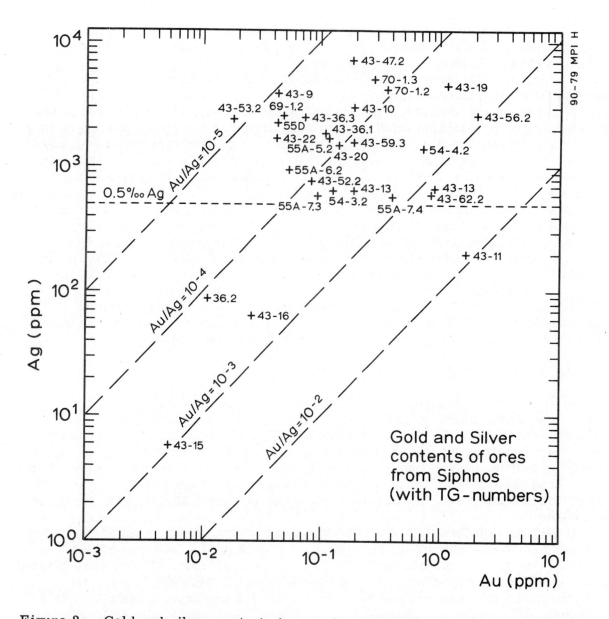

Figure 2: Gold and silver contents in ores from Siphnian mines.

with the silver content of the Laurion galena ores. From the chemical analyses it becomes clear that the ancient mines lying on this mineralization belt must have been lead and silver mines. The low gold contents in the ores make it very unlikely that gold was ever produced from them.

However, preliminary chemical data on some ore samples collected in the area of the Ayos Ioannis (figure 1) iron pits near Faros show besides silver contents up to 0.15% also relatively high gold contents up to 13 ppm. Although it is not yet clear if the Ayos Ioannis site was worked in antiquity, the finds of scattered obsidian flakes and an archaic pot-sherd on the surface between the modern pits may indicate ancient working at this site. At present it seems not unreasonable to assume that the Siphnian gold mentioned in the ancient reports originated from the Ayos Ioannis area and not from the lead-silver mines along the main mineralization belt. Detailed geological and archaeological studies are planned in order to solve these questions.

69

LEAD ISOTOPE ANALYSES

At present lead isotope analyses have been made on samples from only two of the ancient mining sites on Siphnos, Ayos Sostis and Kapsalos-Frase. Samples have also been analysed from the ancient smelting site of Plati Yialos (figure 1). The materials from Ayos Sostis (TG 43) which were analysed for lead isotopic composition include litharge, galena, cerussite and lead antimony sulphosalts. The litharge samples and one galena sample were found strewn over the peninsula to the North East of the church (see topographic map in Wagner and Weisgerber, 1979). The cerussite and the sulphosalts are from below ground and are described in table 1. On the wind-exposed slope and ridge over the Kapsalos-Frase (TG 55A) mine heavy slag and litharge were collected. Nine fragments of litharge from this site were analysed. The lead isotopic composition was also determined on heavy slag and on litharge from the Plati Yialos (TG 53) site. The results of lead isotope analyses for these samples are given in table 3 and plotted in figure 3. It is clear that there is a characteristic field of isotopic composition for the Siphnian ores, slags and litharge so far analysed. This field is both smaller in extent and quite distinct from the Laurion field (figure 4). The Siphnian field is also quite distinct from the isotopic composition of lead in galena from any of the Cycladic islands so far sampled and from all important mainland ancient mining regions around the Mediterranean (Gale 1979a, 1979b).

Table 3: Lead isotopic compositions for ores, litharge and slags from Siphnos.

Sample No.	Description	$^{208}Pb/^{206}Pb$	$^{207}Pb/^{206}Pb$	$^{206}Pb/^{204}Pb$
TG 43-4	litharge	2.07784	.83798	18.701
TG 43-6	litharge	2.07832	.83827	18.714
TG 43-7	litharge	2.07988	.83867	18.697
TG 43-8	litharge	2.07657	.83800	18.746
TG 43-9	galena	2.07945	.83831	18.714
TG 43-10	cerussite	2.08010	.83819	18.751
TG 43-36	green and yellow secondary minerals	2.07844	.83814	18.743
TG 53-1-A	heavy slag	2.07648	.83793	18.712
TG 53-1-B	heavy slag	2.07511	.83774	18.690
TG 53-2-A	litharge	2.08309	.83929	18.755
TG 53-2-B	litharge	2.07804	.83814	18.709
TG 55 A-14	litharge	2.07970	.83829	18.723
TG 55 A-15	litharge	2.07993	.83849	18.743
TG 55 A-16	litharge	2.07757	.83797	18.735
TG 55 A-17	litharge	2.08239	.83845	18.762
TG 55 A-19	litharge	2.07942	.83851	18.775
TG 55 A-20	litharge	2.08060	.83827	18.752
TG 55 A-21	litharge	2.08187	.83838	18.763
TG 55 A-22	litharge	2.08238	.83827	18.757
TG 55 A-23	litharge	2.07978	.83849	18.732

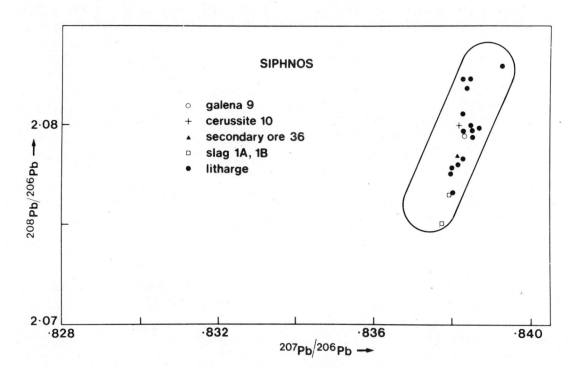

Figure 3:   Lead isotopic composition in Siphnian ores, slags and litharge.

TRACES OF ANCIENT WORKINGS

In spite of intensive destruction by the late 19th and early 20th century iron quarrying numerous traces of the ancient mining and smelting activities still remain on Siphnos, and were reviewed briefly by Gale (1979b).   In the following discussion some further observations made at the sites with ancient workings are briefly reported.   The finds and observations will be described in more detail by Gropengiesser and Wagner (in preparation).

a) Ayos Sostis (TG 43)

This site has been investigated to some degree in co-operation with mining archaeologists from the Deutsches Bergbaumuseum, Bochum, during the summer 1977 expedition.   The results of this investigation and a topographic map of the site were recently published (Wagner and Weisgerber, 1979).   Traces of ancient activities extend over most of the Ayos Sostis peninsula.   Some of them had already been observed by others, especially Bent, (1885) and Graindor (1903).   On the surface scattered slags and ceramic fragments occur frequently.   They seem to be concentrated on the northern slope near the ridge of the peninsula and probably represent remains of ancient smelting furnaces.   Occasionally also fragments of litharge can be found there.   Ancient dumps with rock and ore flakes chipped off by hammer-and-sledge work still exist abundantly.   They usually contain pottery frag-ments and more rarely obsidian flakes.   A fine arrow-head of obsidian (plate 1) was found on the surface of the peninsula, and several crushing stones and stone hammers (plate 2) were discovered.   The ancient galleries can be entered through ancient shaft entrances (one shaft of mine 1 (Wagner and Weisgerber, 1979) has steps carved in its side) and through the modern iron pits.   The ancient galleries of 'mine 1' along

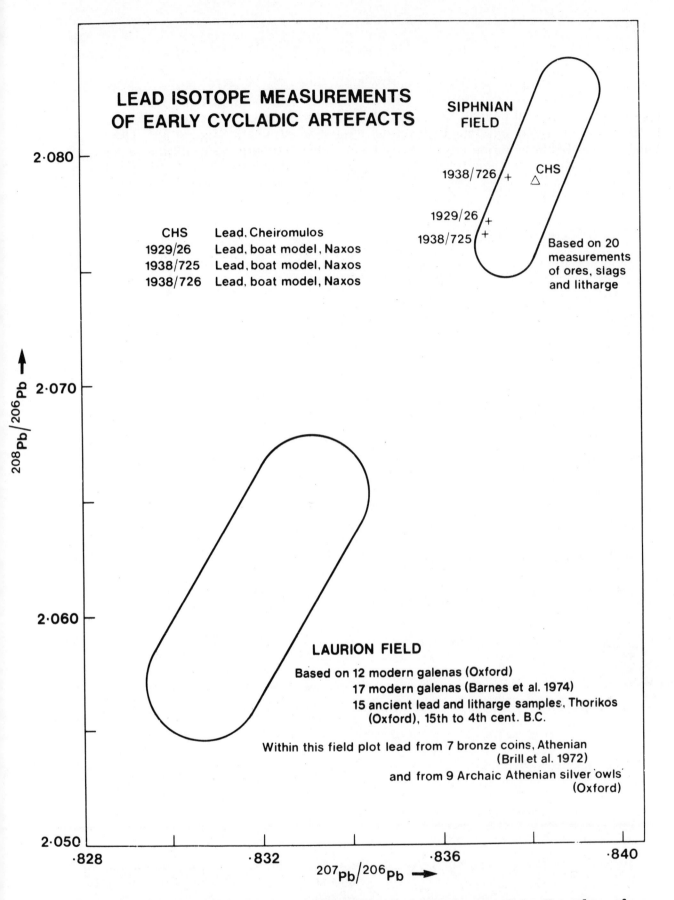

Figure 4: Lead isotope measurements of Early Cycladic objects. Samples of Laurion composition whose lead isotope compositions were measured at Oxford are so designated. The archaic Athenian silver 'owl' tetradrachms are from the Asyut Hoard collection held in Heidelberg.

the vein-like ore body are flooded in their deeper parts to a depth of at least 2.5m.
The water in the galleries tastes fresh, its level is at sea-level and changes with
the sea-tides. This implies a connection between the ancient galleries and the
open sea. Commonly the galleries are filled with loose rock and ore waste
('Versatz'), which in part is cemented by calcareous sinter. In one place ('mine 2')
pottery and charcoal was found within the cemented 'Versatz' (figure 5). The
charcoal probably remains from fire-setting, used anciently as a mining technique
(Collins, 1893). Occasionally pottery and stone hammers can be found in the loose
'Versatz'. On the walls of the ancient galleries usually tool marks of various
kinds (e.g. plate 3a) and rarely lamp recesses are visible.

b) Ayos Silvestros (TG 70)

Near the top of the mountain Ayos Silvestros, about 280 m above the Ayos Sostis
peninsula, modern iron pits intersect several ancient galleries, which are partly
filled with 'Versatz'. The ancient galleries exhibit tool marks on their walls.
Above ground one shaft entrance was discovered underneath brush-wood.

c) Vorini (TG 54)

In the extensive, modern iron pits of Vorini numerous ancient galleries are
exposed, of which plate 4 is a specially instructive example. Some of the galleries
are still filled with ancient 'Versatz' (plate 5). Tool marks are common. Above
ground there are some ancient dumps consisting of chipped-off rock and ore flakes.
They contain pot-sherds - among them a fragment of an oil lamp. Scattered obsidian
flakes were observed.

d) Kapsalos (TG 55)

One of the most extensive ancient mining regions must have been the one at
Kapsalos (plate 6) about 200 to 260m above the Kamaron valley. Ancient galleries
are exposed at the modern iron pits of Kapsalos-Frase (TG 55A), Kapsalos-Bolioni
(TG 55B) and Kapsalos-Tsingouras (TG 55D). Above ground, on the Kapsalos ridge
and on the slope towards the Frase pit, scattered fragments of pot-sherds, slags,
litharge and obsidian flakes were found. Four ancient shaft entrances are still
visible near the Frase-pit, one of them with carved steps. 'Versatz' and various
tool marks are common. Replicas made from tool impressions in ancient galleries
at Kapsalos-Frase are shown in plate 3b; one type seems to belong to axe-adzes,
known as mining implements in various periods, another belongs perhaps to a thick
and pointed chisel-like instrument, possibly of antler or bone. At Kapsalos-Frase
the ancient galleries extend even into slope gravel. The walls of the modern
Kapsalos-Bolioni iron pit intersect numerous ancient galleries, giving the appearance
of a Swiss cheese. At this site good examples of neatly built 'Versatz' walls and
lamp recesses can be observed underground. Some of the 'Versatz' is cemented by
calcareous sinter. One pot sherd was found lying on top of the 'Versatz'. The
ancient galleries at the Kapsalos-Tsingouras site are largely destroyed by the
modern iron mining.

e) Xeroxylon (TG 69)

The ancient workings at Xeroxylon on the ridge of Oros Profitis Elias at about
600m above sea-level seem also to be largely destroyed. Only two filled-in shaft
openings with carved steps and tool marks were seen.

f) Plati Yialos (TG 53)

At the cliff bordering the Plati Yialos bay in the west, following Fiedler's 1840
description, scattered fragments of litharge, heavy slags and abundant obsidian

flakes were found. No traces of mining activities were observed at this site.

g) Ayos Ioannis (TG 71)

At the modern iron pits at Ayos Ioannis near Faros no ancient galleries have yet been observed. However, the occurrence of numerous obsidian artefacts and a fragment of an archaic pot-sherd above ground between the pits may indicate ancient workings at this site.

Taken together these observations imply that in antiquity extensive lead-silver mining and cupellation was practised at various places on Siphnos Island. Of course, one may immediately ask if all these workings were operated simultaneously and over which period they were in use. Therefore, one of the most important aims was the dating of these ancient workings.

DATING

For dating the ancient mines, various materials and objects, such as charcoal, pot-sherds, mining implements and obsidian artefacts may be used. Some of these materials were found at most of the ancient sites. However, at Ayos Sostis all these materials can still be found, and in greater abundance than at the other sites. Therefore, the dating efforts were concentrated at this site. Three independent dating methods were used, namely radiocarbon (C-14), thermoluminescence (TL) and archaeological dating.

In one fortunate case - the 'Versatz' cemented by calcareous sinter in 'mine 2' - all three methods were applied to the same context, which is schematically illustrated in figure 5. Directly on the floor of the gallery rests a layer consisting of charcoal and chipped-off rock flakes tightly cemented together. Within this layer two pot-sherds (samples K 196A and K 196B, plate 7) were found. The charcoal was dated by C-14, the pot-sherds were dated by TL and by archaeological criteria.

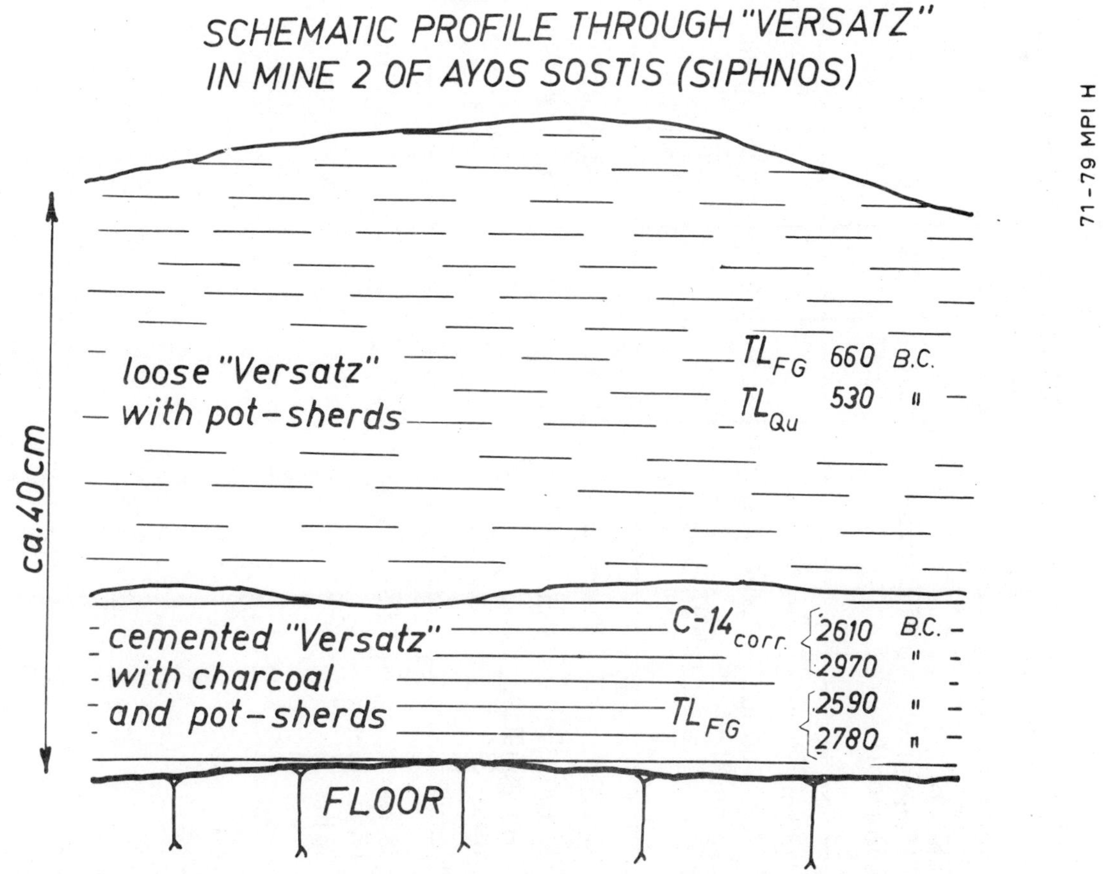

Figure 5: Schematic profile through "Versatz" in mine 2 of Ayos Sostis (context).

Table 4: TL data on pot-sherds from ancient galleries on Siphnos

Sample	TL-technique	dose (in rad)				annual dose-rates (in rad/y)				age (in y)	time	error (in y)	
		ED_β	$+ I_o =$ AD	a	a_{eff}	$a_{eff} +$	$\beta +$	$\gamma =$	total			random	total
K 194 (Kapsalos-Bolioni)	Fine-grain	476	128 604	0.107	0.111	0.111	0.227	0.028	0.366	1649	330 A.D.	± 50	±140
K 195 (Ay. Sostis, mine 2)	Fine-grain	832	197 1029	0.202	0.196	0.196	0.172	0.023	0.390	2635	660 B.C.	±100	±220
	Quartz	446	– 446	–	–	–	0.155	0.023	0.178	2511	530 B.C.	±140	±230
K 196A (Ay. Sostis, mine 2)	Fine-grain	1033	– 1033	0.141	0.070	0.070	0.133	0.022	0.226	4571	2590 B.C.	±150	±320
K 196B	Fine-grain	968	– 968	0.233	0.071	0.071	0.110	0.022	0.203	4760	2780 B.C.	±150	±380
K 198 (Ay. Sostis, mine 2)	Fine-grain	757	150 907	0.109	0.110	0.110	0.179	0.024	0.313	2900	920 B.C.	±100	±250
	Quartz	405	91 496	–	–	–	0.161	0.024	0.185	2682	700 B.C.	±160	±250

For C-14 dating two samples of charcoal (samples H 5048-4469 and H 5471-4015) were collected and submitted to Professor O. Münnich of the C-14 Laboratory, Institut für Umweltphysik der Universität Heidelberg. The conventional C-14 ages were determined as 4250 ± 180 years BP (H 5048-4469) and 4000 ± 50 years BP (H 5471-5015). The ages were calculated with the 5568 y half-life and were corrected for isotope fractionation; the errors are at the 1σ level. Dendrochronological calibration, using the MASCA conversion table by Ralph, Michael and Han (1973), places these ages about 2970 B.C. and 2610 B.C. respectively. For TL dating the two pot-sherds were washed in dilute acetic acid in order to remove the calcerous sinter. The fine-grain and coarse-quartz fractions were separated. Due to un-favourable TL-properties (no plateau, shifting of the glow-curve peak, strong supralinearity) the quartz fractions were not used for dating. The fine-grain frac-tions behaved well in terms of their TL properties (plateaus within 10% above 350°C glow-curve temperature, no anomalous fading above 400°C after 6 weeks storage time). The dose-rate was determined by α-counting, fission tracks, neutron-activation, atomic absorption and on-site γ-scintillation counting. The dose-rates were corrected for water content. Detailed data are given in table 4. The cal-culated TL fine-grain ages are 2590 B.C. ± 320 y (sample K 196A) and 2780 B.C. ± 380 y (sample K 196B). The errors are the total analytical 1-σ errors. The two sherds K 196A and K 196B were also dated using archaeological methods. Both fragments are of handmade coarse ware. Fabric, techniques and shape of the vessels are characterised in table 5. According to these criteria both fragments are of Early Bronze Age (EBA): K 196B seems to belong to an early stage of EBA, K 196A seems to have some Neolithic features. All three independent dating methods applied to the finds from the same context in mine 2 appear to agree with each other rather well if we assume chronologies which place the beginning of the EBA in the first half of the third millenium.

TL dating was also applied to single fragments of some other pot-sherds in mine 2 (plate 7): sample K 195 from the overlying loose 'Versatz' (figure 5) and sample K 198 from another occurrence of 'Versatz'. Their data are given in table 4. Since the latter TL ages are measured on single samples they must not be over-interpreted. The archaeological ages of these pot-sherds are of very similar early date to K 196A and K 196B. Both are hand-made coarse ware and their clay con-sistence resembles that of K 196A and K 196B. Their fabric, techniques and shape are described in table 5. More detailed suggestions about the likely different archaeological stages represented by these four coarse ware sherds of mine 2 cannot be made, because insufficient stratified ceramics of this kind from the Cyclades are yet published.

However, some pot-sherds collected above ground at Ayos Sostis, e.g. KAS 1.78 and KAS 3.78 (plate 8), can be more accurately dated on the basis of their fabric, techniques and ornament (table 5). The practise of mat-impression on fragment KAS 1.78, hitherto unknown from Siphnos, was in favour in the Cyclades during the Late Neolithic and the EBA. The fabric of the sherd also suggests an early stage. Fragment KAS 3.78 with the herring-bone ornament belongs to the Grotta-Pelos Culture of Early Cycladic I.

Apart from these pre-historic pot-sherds later ones were also found at Ayos Sostis; many of them are of Late Archaic date. The sherds observed at the other ancient mining sites are mostly of Late Archaic age. So far, no pre-historic sherd has been identified at these sites. The wheel-made fragment K 194 (compare table 5), which was found lying on the 'Versatz' inside Kapsalos-Bolioni, is archaeo-logically not yet dated. Its TL age is given in table 4.

The other finds at Ayos Sostis become comprehensible in terms of the pre-historic age now established for the mine at this site. The obsidian arrow-head (plate 1) belongs to the Late Neolithic tanged type which is also reported from EBA sites, at least on the Greek mainland. The obsidian flakes are probably remains of the pre-historic fabrication of stone implements, used by the miners. There is some doubt about the type and material of the chisel-like tool, the marks of which were observed in the walls of the mine (plate 3a). The slightly incurving surface of

Table 5: Fabric, techniques, shape and ornament of pot-sherds.

Ayos Sostis, mine 2:

K 196A : rim fragment of a thickwalled large bowl (diam. ca. 0.24m) with plain and full-rounded lip; handmade; relatively rough and very micaceous clay with particles of quartz grains and of micaschist and with brick inclusions; reddish brown in colour, not very hard fired and gritty to the touch; the rough looking surfaces originally smoothed; on the outer surface barely visible traces of scoring.

K 196B : wall fragment, probably from a thick-walled jar; handmade; brown to orange-brown clay with gray core, consistence and firing like K 196A; on the outer surface a thin smooth clay layer, somewhat harder and greyish-brown fired, with shallow and crossing scores in various directions; on the inner surface deeper scores partly parallel.

K 195 : rim fragment of a thickwalled deep bowl (diam. ca. 0.30m); handmade; orange-brown clay, consistence as K 196A, but relatively well fired and of somewhat rougher fabric; on the outer surface irregularities, especially in the left part and at the rim, also traces of thin black slip; the fully rounded lip partly thickened on the outside; the inner surface completely lost.

K 198 : rim fragment of a thickwalled bowl (diam. ca. 0.30m) with a plain full-rounded lip as K 196A, in shape somewhat deeper; orange-brown clay with greyish core, consistence and firing as K 196A; the surfaces smoothed; remains of scores especially on the inner surface, where they run diagonally; traces of thin black slip, the so called "Urfirnis" on both surfaces and on the lip.

Ayos Sostis, above-ground:

KAS 1.78 : fragment of base with mat-impression (diam. 0.071 : 0.051m) from a thick walled vessel of uncertain shape; handmade; very rough, red clay, partly brown, with much fine and coarse mica, large and small inclusions of quartz grains and mica-schist; surfaces very rough.

KAS 3.78 : fragment with herring-bone ornament from a thickwalled vessel of uncertain shape; handmade; very rough clay with much fine mica, inclusions of quartz grains and large particles of grey mica-schist; redbrown to orange in colour with the characteristic Cycladic purple gleam.

Kapsalos-Bolioni :

K 194 : base (diam. 0.093 : 0.089m) of a relatively thinwalled vessel, probably of a jar; wheelmade and fired hard; red-brown clay, the inner part of the wall fired greyish; many mica and quartz inclusions; the outer surface spotted, greyish.

the broader side, still visible at least on the cast bottom left of plate 3a, and the rounded edges might suggest a tool of bone or antler (one similar cast was also obtained from the walls of Kapsalos-Bolioni); on the other hand the wear depressions in the stone hammers and the tool marks on the harder rock faces inside the mines suggest the use of metal chisels or picks (the use of antler or bone tools is suggested at Rudna Glava, see page 36, Ed). In any case the use of six-sided stone-hammers (e.g. plate 2a) and the triangular 'crushing stones' or celts (e.g. plate 2b) must originate from the first miners. This class of implements is traditional for much older periods and is reported to have been used elsewhere in Neolithic times also by copper miners. (Forbes, 1950 Davies, 1935). However stone implements like these from the Ayos Sostis site were also in use during the Early Bronze Age; they occur in corresponding types in many settlements and at burial sites, also in the Cyclades. Possibly they were used by miners also in later, even Archaic times, when mining continued on the site or was practised again.

SOURCE OF OBSIDIAN ARTEFACTS

Obsidian flakes and artefacts are of widespread occurrence on Siphnos near the mining sites. Obsidian flakes are especially common at the Plati Yialos smelting and cupellation site. At the 1979 Symposium on Archaeometry in London there was described a method for characterizing obsidian sources around the Mediterranean based on very precise strontium isotope analyses and precise rubidium and strontium analyses (Gale, 1979c). This technique easily distinguishes the obsidian sources on Melos, Giali, Antiparos, Pantelleria, Sardinia, Palmarola and Lipari. Moreover it can easily distinguish between the two sources on Melos at Dhemenegaki and Sta Nychia. Obsidian artefacts from Ayos Sostis and from Plati Yialos have been characterized by this method; in both cases all artefacts analysed prove to have come from the Dhemenegaki source on Melos.

This result was confirmed independently by fission track uranium analysis. Geological occurrences of obsidians at Melos, Chios, Santorini, Giali, Lipari and several Anatolian sites have been characterized by the uranium content. Most of these occurrences can be distinguished by their uranium contents. For fourteen obsidian flakes and artefacts from Siphnos uranium contents ranging between 2.55 and 3.42 ppm were determined which match very well the low uranium content of the Melos source (Wagner, in preparation).

SUMMARY AND CONCLUSIONS

Field and laboratory studies have revealed that the ancient Siphnian mines at Ayos Sostis, Ayos Silvestros, Vorini, Kapsalos and Xeroxylon were worked chiefly for complex lead-antimony-silver ores, exploiting them both for lead and silver. The chemical and isotope studies suggest that these ore deposits are all in the iron hats (gossans) of a single sulphide mineralization at great depth. Gold was not detected in any minable concentration in these mines. However relatively high gold contents were discovered in the ores from Ayos Ioannis iron pit, suggesting that the ancient gold mines may more probably be found in that region.

Heavy slags and litharge - associated with ancient pot-sherds and obsidian fragments - were found at Ayos Sostis, Kapsalos and Plati Yialos. The finds of litharge provide evidence that cupellation was practised on Siphnos in antiquity.

Radiocarbon and thermoluminescence dating place the ancient workings at Ayos Sostis in the first half of the third millenium. According to archaeological observations made at this site lead and silver production started on Siphnos at least in Early Cycladic I. This is the first proof for lead and silver mining and cupellation during the Early Bronze Age in the Cyclades (Wagner et al., 1979). There is also evidence for mining at Ayos Sostis in Early Cycladic II and later periods of the Bronze Age, based on pot-sherds which we have not yet published.

Gale (1978) reported lead isotope data on three lead boat models found on Naxos in an Early Cycladic II context, and a cast lead sample from Cheiromylos found on the surface at the Grotta-Pelos EC I settlement site in Dhespotikon (Renfrew, 1967). These objects have isotopic compositions which overlap the Siphnian field (figure 4). Assuming that no other lead sources of similar isotopic composition to Siphnos exist (and none have yet been found) Gale (1978, 1979b) suggested "that lead working on Siphnos may have a very early beginning indeed in EC II .... times". In fact the sample from Cheiromylos indicates lead working on Siphnos from EC I times. This prediction is now found to be in excellent agreement with the present direct age determinations of the Siphnian mining activities.

Silver mining on Siphnos was also practised in the Late Archaic period, as was mentioned by Herodotus and is now confirmed for the Ayos Sostis, Vorini and Kapsalos sites. It might, therefore, have been a source of silver for some Archaic Greek coins. Work reported elsewhere (Gale, Gentner and Wagner, 1980) shows this to be true.

The evidence for lead and silver working on Siphnos already in Early Cycladic I has the important implication that lead-silver mining and metallurgy was practised in the Cycladic region in these early times and that there was no necessity to import these metals from elsewhere.

Acknowledgements

Many of our colleagues contributed to this work by giving advice, submitting samples, co-operating in the field and performing experimental analyses. We would like to thank Dr. H. G. Bachmann (Hanau), I. Bassiakos (Athens), M. Bruns (Heidelberg), Dr. H. Catling, Dr. R. Jones, Dr. C. Mee (Athens), Professor St. Charalambous (Thessaloniki), BA H. G. Conrad (Bochum), Dr. B. Dominik, Professor A. El Goresy, A. Haidmann (Heidelberg), Professor Sp. Jakovidis (Philadelphia), Professor N. Jalouris (Athens), D. Kaether, E. Klein (Heidelberg), Professor K. Kilian (Athens), C. Lehr, Professor O. Münnich, K. Oberfrank, R. Pelikan (Heidelberg), Dr. P. R. S. Moorey, Dr. H. J. Case, Dr. M. Vickers (Oxford), Professor H. Mussche, Dr. P. Spitaels (Ghent), Dr. E. Pernicka (Vienna), Dr. B. Philippaki, Professor J. and E. Sakellarakis (Athens), Professor A. C. Renfrew (Southampton), U. Schwan (Heidelberg), Dr. J. Stavropodis (Athens), Dr. S. Warren (Bradford), Dr. G. Weisgerber (Bochum), A. Weirich (Heidelberg), Dr. O. Williams-Thorpe (Milton-Keynes), Professor. R. J. Hopper (Sheffield), and J. Birkett-Smith (London).

The project was supported with funds of the Stiftung Volkswagenwerk.

Finally we would like to dedicate this paper to the memory of our colleague Dr. O. Müller (Heidelberg), who remained fully involved in the work until his sad and untimely death on 4 January 1979.

REFERENCES

Barnes, I. L.,   1974.   Isotopic analyses of Laurion lead ores.   In
Shields, W. R.,   Archaeological Chemistry Ed. C. W. Beck,
Murphy, T. J. and   Advances in Chemistry Series 138, American
Brill, R. H.   Chemical Soc., 1-10.

Bent, J. Th.   1885.   On the gold and silver mines of Siphnos. <u>Journal of Hellenic Studies 1885</u>, 195-198.

Brill, R. H. and   1972.   Lead isotopes in ancient coins. <u>Special Publication No. 8, Royal Numismatic Soc.</u>, London, 279-303.
Shields, W. R.

Collins, A. L.   1893.   Fire-setting, the art of mining by fire. <u>Transactions of the Federal Institute of Mining Engineers</u>, V, 82-92.

Davies, O.   1935.   "<u>Roman Mines in Europe</u>", Oxford, pp34-38.

Eustathios: Commentary on Dionysius Periegetes 525, 27ff. In C. Muller, <u>Geographi graeci minores II</u>, 201-407, Paris, Didot, 1861.

Fiedler, K. G.   1840.   Reise durch alle Theile des Königreiches Griechenland im <u>Auftrag der königlich-griechischen Regierung</u>, Leipzig 1840, 125-144.

Forbes, R. J.   1950.   "Metallurgy in Antiquity", pp56-59, Brill, Leiden 1950.

Gale, N. H.   1978   Lead istopes and Aegean metallurgy. Proceedings of the 2nd. <u>International Scientific Congress on Thera and the Aegean World</u>, 529-545.

Gale, N. H.   1979a.   Lead isotopes and archaic Greek coins. <u>Proceedings of the 18th International Symposium on Archaeometry, Archaeo-Physika 10</u>, 194-208.

Gale, N. H.   1979b.   Some aspects of lead and silver mining in the Aegean. <u>Miscellanea Graeca, Fasc. 2</u>, 9-60.

Gale, N. H.   1979c.   Gypsum, marble, obsidian: Can their provenance be determined by strontium isotope analyses? <u>19th International Symposium on Archaeometry, London, Abstract Volume</u>.

Gale, N. H.,   1980.   Mineralogical and Geographical Silver Sources of Archaic Greek Coinage. Royal Numismatic Society, Special Publication No. 12, <u>Metallurgy in Numismatics, Volume I</u>. (In Press).
Gentner, W. and
Wagner, G. A.

Graindor, P.   1903.   Mines anciennes en Grèce. <u>Le Musée Belge VII</u>, 466-470.

Klein, E.   1979.   Chemical and mineralogical studies of Siphnos ores and slags. <u>Archaeophysika 10</u>, 223-229.

Müller, O. and   1979.   On the composition and silver sources of Aeginetan coins from the Asyut Hoard. <u>Archaeophysika</u>, <u>10</u>, 176-193.
Gentner, W.

Ralph, E. K.,   1973.   Radiocarbon dates and reality. <u>MASCA Newsletter 9</u>, 1-20.
Michael, H. N. and
Han, M. C.

Renfrew, A. C.   1967.   Cycladic metallurgy and the Aegean Early
                                 Bronze Age. American Journal of Archaeology
                                 67, 1-20.

Wagner, G. A.,   1979.   Evidence for third millenium lead-silver mining
Gentner, W. and   on Siphnos island (Cyclades). Naturwissen-
Gropengiesser, H.   schaften 66, 157-158.

Wagner, G. A. and   1979.   The ancient silver mine at Ayos Sostis on
Weisgerber, G.   Siphnos (Greece). Archaeophysika 10, 209-222.

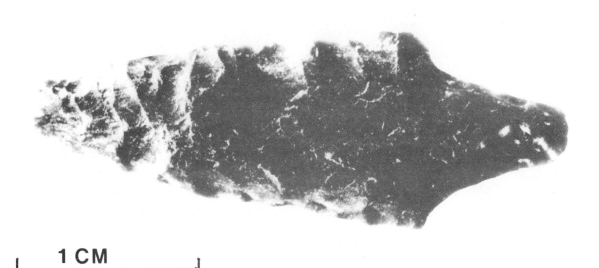

1 CM

Plate 1: Tanged arrow-head of obsidian from Ayos Sostis, found above-ground.

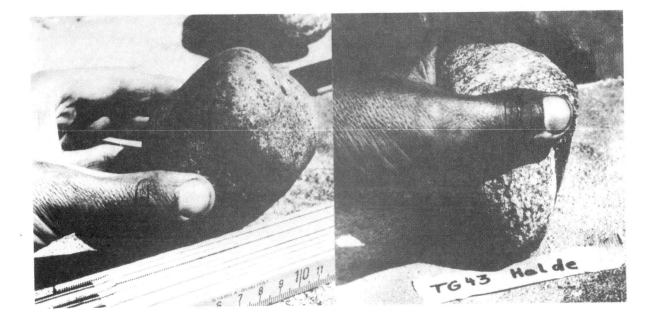

Plate 2: Stone implements from Ayos Sostis mine:
    a) stone hammer
    b) celt ("crushing stone")

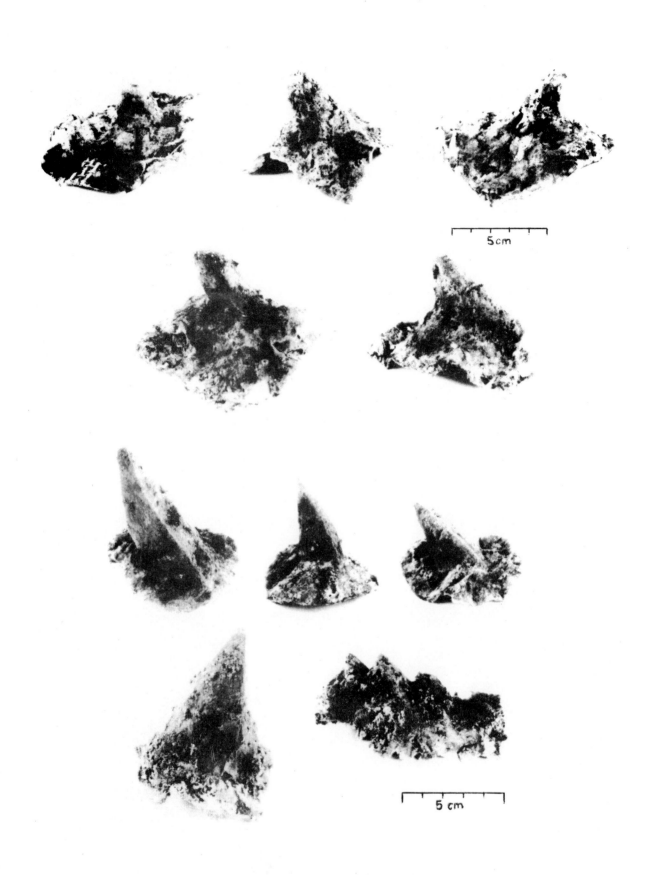

Plate 3: Casts of tool impressions on gallery walls:
       a) from Ayos Sostis
       b) from Kapsalos-Frase

TG 54 Vorini (Siphnos)

ca 10 m

Plate 4: Ancient gallery system vertically intersected by modern iron pit Vorini.

ca 2 m

Plate 5: Cross section of ancient galleries exposed at wall of modern iron pit Vorini.

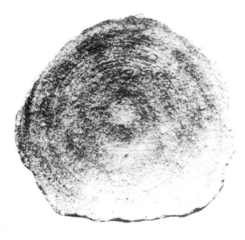

K 194   2 cm

K 194   2 cm

K 195   2 cm

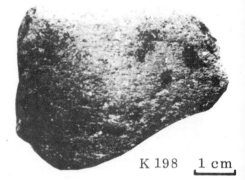

K 198   1 cm

K 196 A   1 cm

K 196 B   2 cm

Plate 6:   Dated pot-sherds from galleries: K 194 from Kapsalos-Bolioni;
K 195, K 198, K 196A and K 196B from Ayos Sostis mine 2.

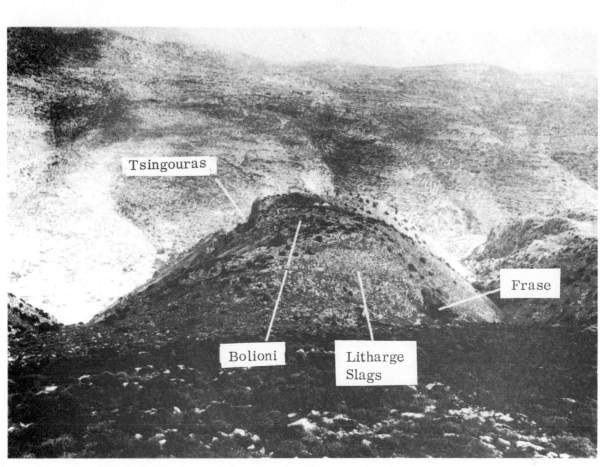

Plate 7: The Kapsalos-ridge with mining sites.

Plate 8: Pot-sherds KAS 1.78 and KAS 3.78 from Ayos Sostis above-ground.

DATA ANALYSIS: TOWARDS A MODEL OF CHEMICAL MODIFICATION OF COPPER FROM ORES TO METAL

Th. BERTHOUD, S. BONNEFOUS, M. DECHOUX, J. FRANÇAIX

Commissariat à l'Energie Atomique - Centre d'Études Nucléaires de Fontenay - aux - roses - SEA/SCAA - BP No 6 92260 FONTENAY AUX ROSES.

Summary

Two series of minor and trace element data, one on copper ores, and the other on copper objects are the basis of a quantitative determination of the modification of copper through the steps of the production of copper metal. A third series of data on slags brings precision to these steps. It is then possible to apply the inverse transformation and get from the objects data a "virtual" composition of the original ores. The comparison with the real data on copper ores allows a precise determination of the origin of the copper in objects if it originates from clearly different geochemical contexts. These techniques bring new light on some aspects of the Middle East copper trade in the IVth and IIInd millenium B.C.

Keywords: COPPER, METALLURGY, MIDDLE EAST, MODEL, MULTIVARIATES, STATISTICS, ORES, IRAN, PERSIAN GULF, MESOPOTAMIA TRACE ELEMENTS, TRADE, PROVENANCE STUDIES, ANALYSIS.

INTRODUCTION

This paper presents the final step of a research programme started five years ago by the "Commissariat à l'Energie Atomique" in cooperation with the "Centre National de la Recherche Scientifique - RCP 442", the "Laboratoire de Recherche des Musées de France", and the Universities of Paris I and Paris VI, on the application of spark source mass spectrometry to the study of ancient copper metallurgy. (Berthoud 1977, 1979, and Berthoud et al 1978a).

We have attempted to make a comparison between two series of data, one on copper objects, the other on copper ores. The purpose of such a work is, obviously, to see if it is possible to find the original ores which were used to make the copper in the objects.

Figure 1 shows the area and the archaeological sites concerned with this study. The sites have produced the evidence of IVth and IIIrd millenium BC metallurgy which is the basis of our work. Most of the objects come from precise stratigraphical positions.

Figure 2 represents the mining zones which have been sampled. All the mines show remains of ancient mining. (Berthoud et al 1975, 76, a & b, 77, and 78b).

On the objects and ores, two analytical techniques were used for cross control and calibration of the results:

- emission spectroscopy for major and minor elements

- spark source mass spectrometry for minor and trace elements.

Usually 40 elements were measured, going up to 49 in the ores. The detection level is always below 0.1 ppm.

Independently, the two series of data were treated by different statistical methods : finally principal component analysis seems to be the most convenient techniques for the type of data we get (especially suited to the very wide range of concentration of most of the elements).

Figure 3 shows the plot generated by the 2nd and 3rd principal components on the objects. Twenty four elements have been introduced, although As, Sn, Sb, Pb, Sn, Cd, Bi were not taken into account because it appears in the steps described below that these elements are not usable for determination of the sources of ores. These elements only give technical indications of the way the objects were made. On the use of arsenic it appears that some objects have a low level of arsenic, some have high levels due to the ores (for example the high level of arsenic in the native copper of Talmessi, (Berthoud 1979), and some others have high levels due to an addition of arsenic to make the melted copper more fluid and easier to cast.

Figure 3 shows a clear differentiation between the various groups of chronologically or geographically defined material. Particularly interesting is the close over-lap of Susa B (Vase à la Cachette) and Umm-An-Nar objects.

Figure 4 is also produced by statistical treatment with principal component analysis. It shows the plot generated by component 1 and 3 based on the ore data. This particular feature indicates clear differences between Omani and Iranian copper ores. Ni, Co, V, Cr are clearly in higher concentration in the Omani copper ores than in the Iranian ores. On other plot it is possible to distinguish between the various Iranian ores. An interpretation of all these results is possible by making connections between geochemical zones and geological structures.

With these two series of data our problem was to find a way to establish a correspondence between ores and objects which could allow their comparison and give a way of determination of the ores used in the objects.

There were many possibilities; from them we chose one of the simplest. Our way, was to make a correspondence, element by element, of the overall distributions of ores and objects. Due to certain characteristics of the principal component analysis we established our correspondence by a linear transformation on average and standard deviation. As we used logarithmic scales it gives the following equation:

for each chemical element :

$$Q_{M_V} = \frac{M_M}{Mo^{\; \delta_M/\delta_o}} \; x \; (Qo)^{\delta_M/\delta_o}$$

where

Q_{M_V} : concentration of the element in the 'virtual' ores which should have been in the original ore of the copper in the object

Qo : concentration of the element in the object considered

M_M : geometrical average of the ores distribution

Mo : geometrical average of the objects distribution

δ_M : standard deviation of the ores log distribution

δ_o : standard deviation of the objects log distribution

Through this equation we get a set of data of "virtual ores" which have the same average and standard deviation as our data on real ores and which give in a principal component analysis treatment exactly the same results as the real data on

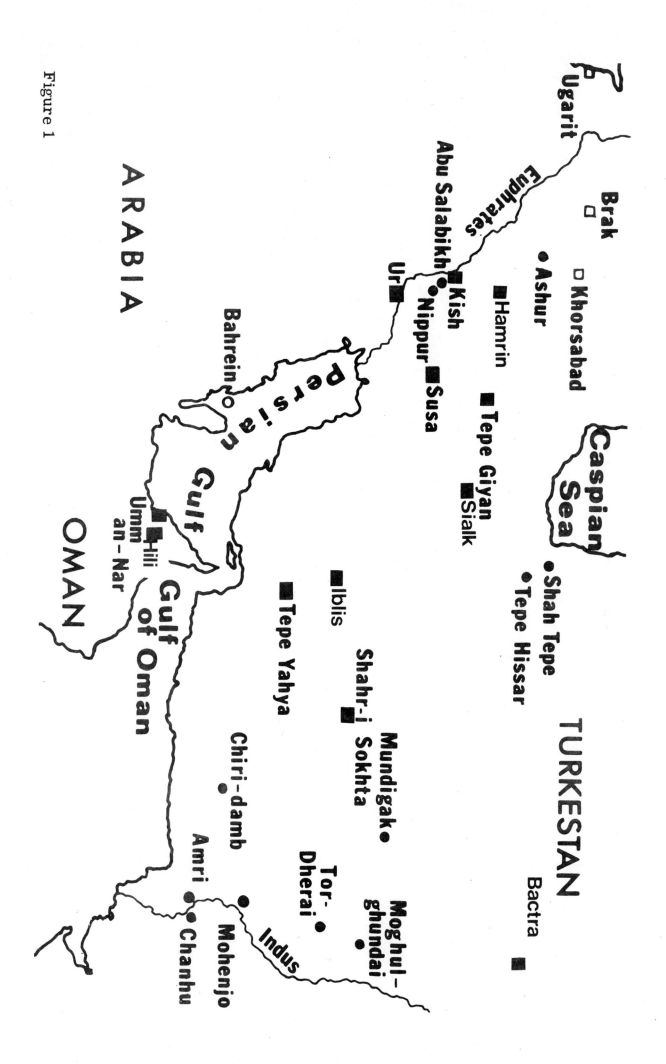

Figure 1

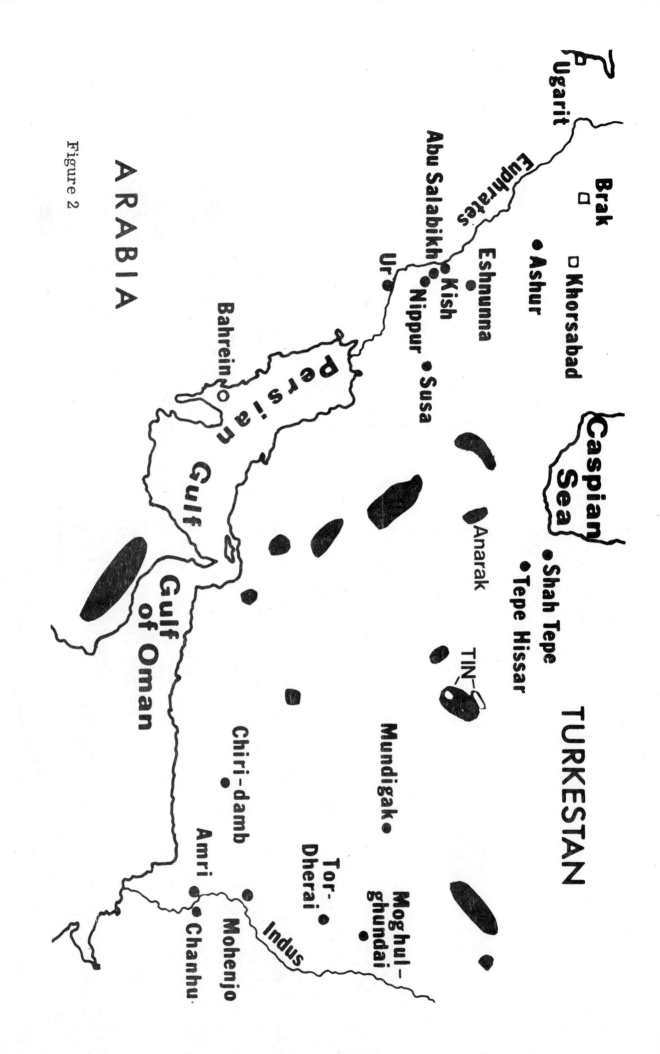

Figure 2

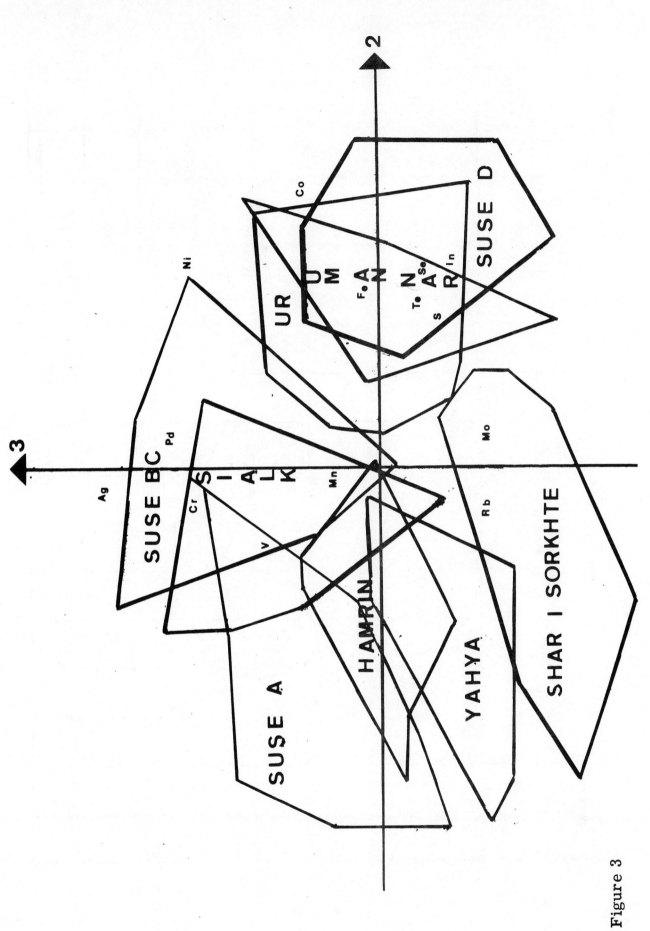

Figure 3

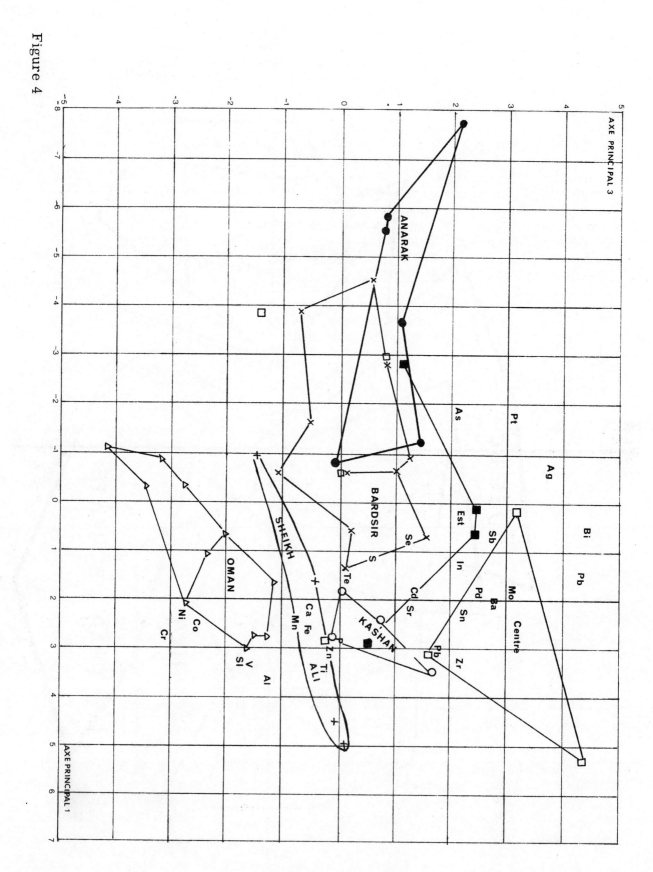

Figure 4

objects. The total information obtained on the objects is entirely retained on transformation.

Figure 5, 6 and 7 show the effect of this transformation on 3 elements (Cr, Mn, Ni). On each of these figures the upper histogram shows the real measurements of the ores (hatched) and the distribution of "virtual ores" produced through the transformation. The lower histogram shows the distributions of the ores and objects.

On the lower histograms the averages and standard-deviations of the objects and ores are different; on the upper histograms the averages and standard deviations of ores and "virtual" ores are the same. On these figures we see that the Cr distribution in the ores and the objects is quite the same, while the distribution of Mn in the ores is higher than the distribution in the objects. For Ni, there appears a concentration indicating that, from a chemical point of view, Ni follows Cu.

On the basis of such histograms, it seems that it is not possible to take into account As, Sn, Sb, Pb, Bi for which we find the levels in the objects are too high in comparison to the ores. Looking at the correlation and the values of each element it seems that As should have introduced Sb and Bi. Sn and Pb are known to be added for metallurgical purposes, and Zn should also have introduced Cd.

Figure 8 shows correspondence between ores (ordinate) and objects (abscisa). For each element (only few are drawn) a different line appears. From a qualitative point of view, all these lines correspond to the metallurgically known results, and for some of them there is a good quantitative correspondence with published data (Tylecote 1977).

Table 1 gives the value of the parameters of all the lines for the 24 chemical elements used in the following data treatment.

The two series of data, real ores and virtual ores, may be now compared directly; we have applied principal component analysis to the complete set of data, real and virtual ores. Figure 9 shows the plot generated by the components 1 and 2 of this analysis. The correspondence between ores and objects is simply established by the overlap of the different groups. It appears that:

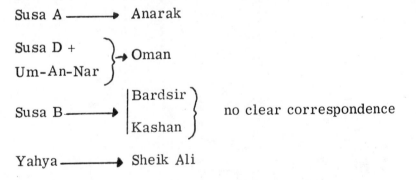

We will not discuss here all the archaeological problems of these results.

The same types of analysis were performed on slags. They were divided in two series:- slags coming from archaeological sites
        - slags coming from mining sites.

Through these analyses it is possible to recognize the two types of slags. The slags from archaeological sites have characteristic high levels of As, Sb, Pb, and Bi, giving a new confirmation of the addition of these elements during one of the final steps of metallurgy. Cleary such an operation did not happen during the first steps of metal production on the mining sites.

From a qualitative point of view, it seems possible to establish a good correlation between the slags and the ores of the same mine but work is only just starting on this problem.

In fact, the way to integrate the slag data in our model is not obvious from either a mathematical point of view or a metallurgical one.

Figures 10 and 11 show the schematic results of copper and tin trade during the IVth millenium (figure 10) and the begining of the third (figure 11). The evidence

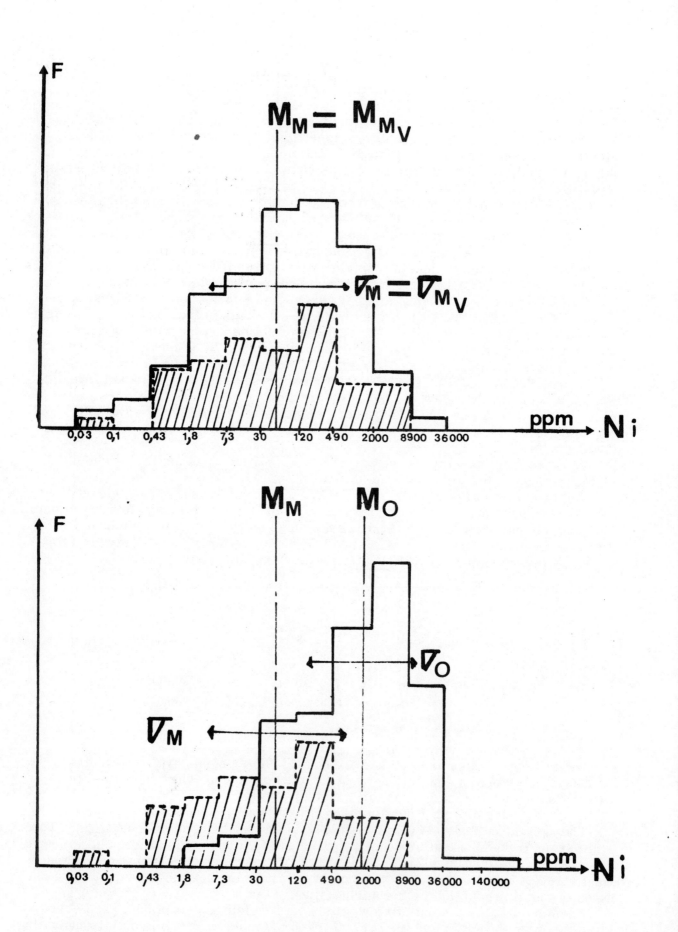

Figure 5 :

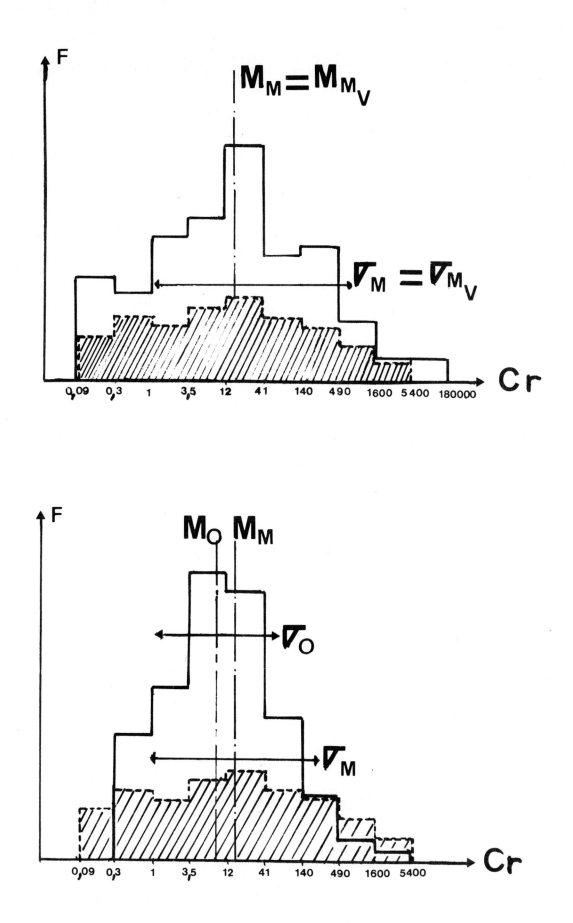

Figure 6 :

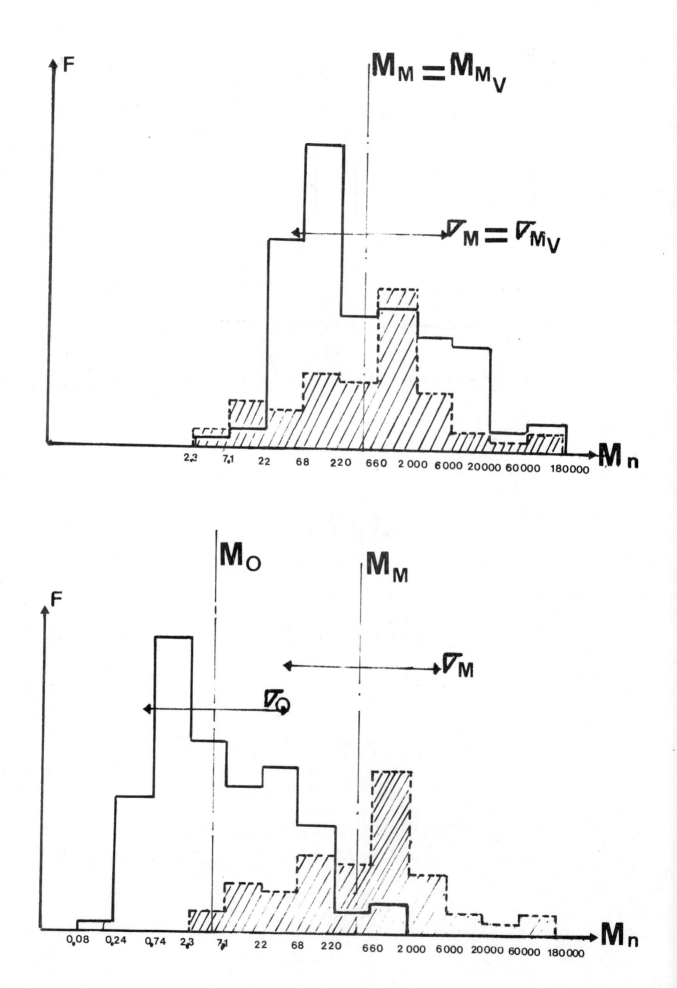

Figure 7 :

Table 1

	M_M	Mo	o_M/oo	$M_M/(M_o \delta m/\delta o)$	M_M/Mo
Al	5660	254	1.217	6.69	22
Si	45000	1622	0.77 2	150	27
P	55	53	1.13 8	0.60	1
S	2500	1105	1.74 8	0.012	2.26
K	1050	572	1.02 3	1.58	1.83
Ca	6073	919	1.19 4	1.75	6.6
Ti	212	18	1.02 3	11.0	11.7
V	24	2.45	0.90 3	10.7	9.8
Cr	14	11	1.40 5	0.48	1.2
Fe	20 500	1667	0.98 9	13.6	12
Mn	443	6.04	1.06 2	65	73
Ni	52	1233	1.29 6	0.0051	0.042
Co	27	57	1.23 5	0.18	0.47
Se	9.28	47	0.80 3	0.42	0.19
Rb	3.9	0.77	0.757	4.75	5.06
Sr	35	11	0.83 9	4.68	3.18
Zr	6	1.56	1.22 9	3.47	3.84
Mo	4.16	1.35	0.84 3	3.23	3.08
Pd	0.68	0.34	1.03 4	2.08	2
Ag	21.9	544	1.46 7	0.0021	0.040
In	1.1	1.76	0.84 8	0.68	0.62
Te	1.32	9.65	0.52 8	0.39	0.13
Ba	11.5	10	1.42 1	0.43	1.15
Pt	0.4	1.82	0.39 8	0.31	0.21

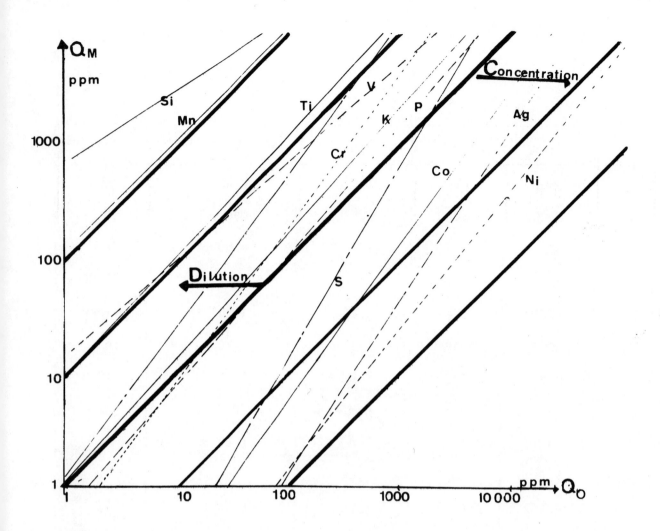

Figure 8

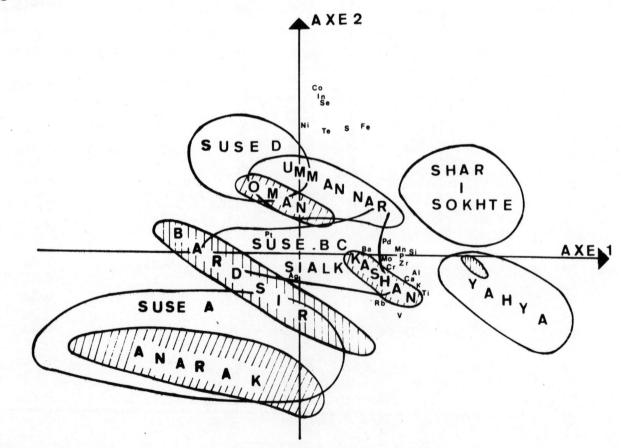

Figure 9

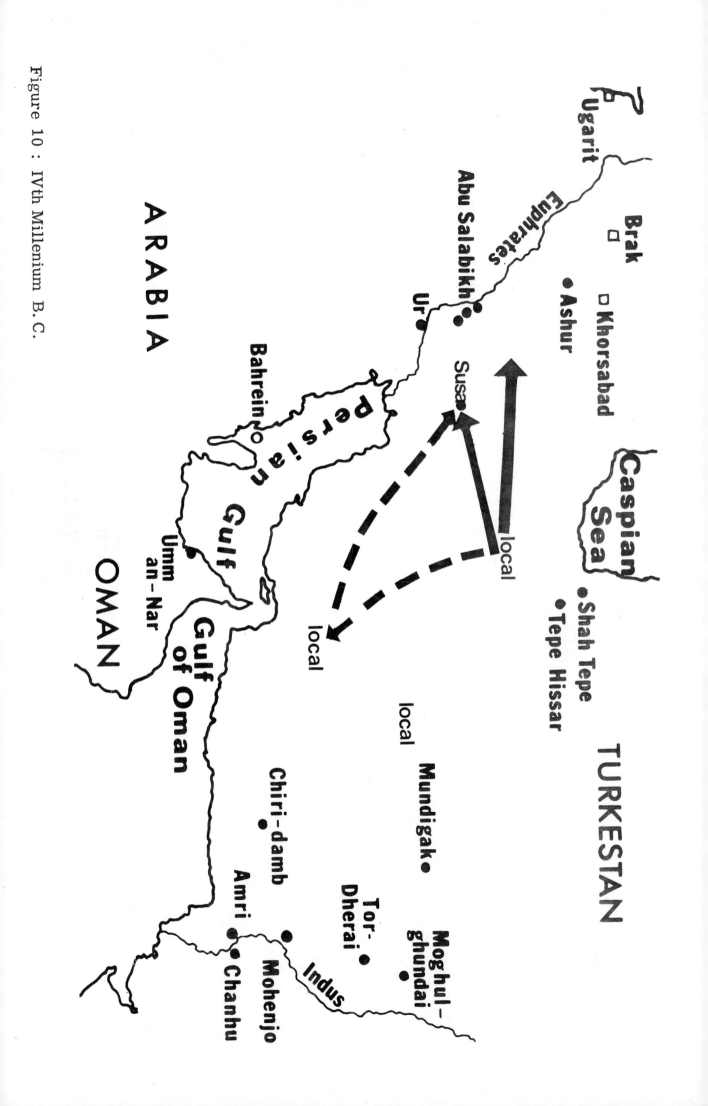

Figure 10 : IVth Millenium B.C.

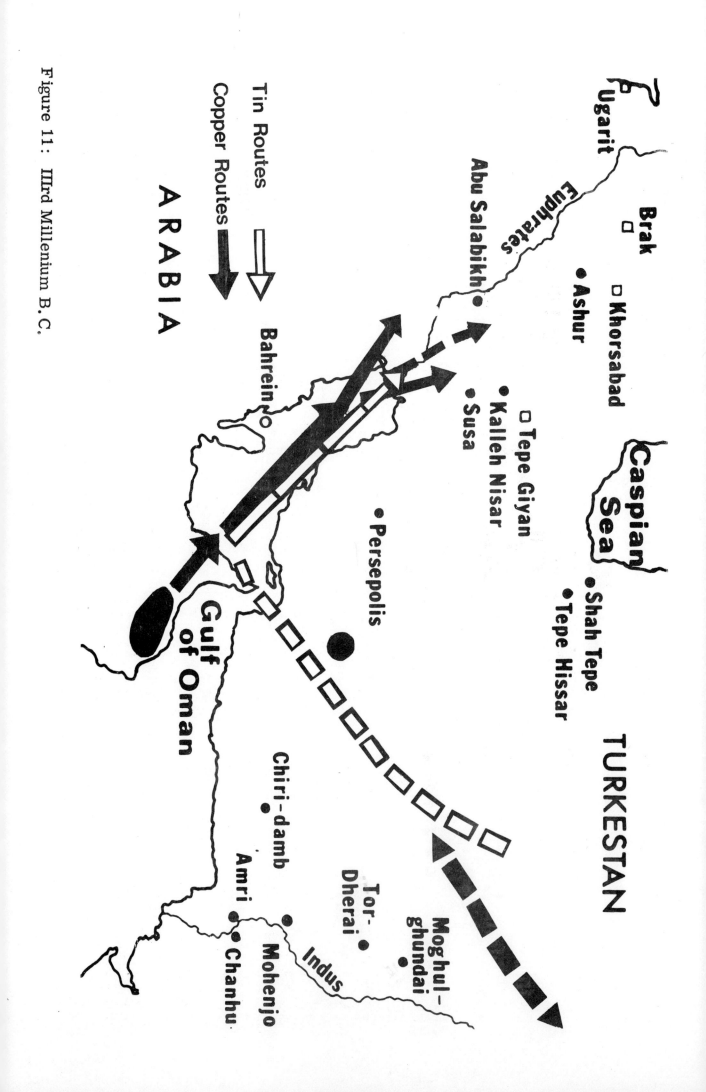

Figure 11: IIIrd Millenium B.C.

for tin trade is not extensive but the few analytical and geological data are in good agreement and are very coherent with respect to the archaeological problems (Muhly 1977).

In conclusion, we would like to point out that our attempt to interpret our analytical results is certainly introducing many problems but it seems to us that both from the archaeological and metallurgical point of view these results indicate that building models of metallurgical transformation may be now considered as a possible and realistic way to study ancient metallurgy.

REFERENCES

Berthoud, T., Besenval, R., Cleuziou, S. and Liszak-Hours, J. 1975. Étude sur la métallurgie iranienne au IVème et IIIème millénaire. Prospection en IRAN - Paris - Multicopie.

Berthoud, T., Besenval, R., Cesbron, F., Cleuziou, S. and Liszak-Hours, J. 1976a. Les anciennes mines de cuivre en IRAN. Deuxième rapport préliminaire. Prospection en IRAN - Paris - Multicopie.

Berthoud, T., Besenval, R. and Cleuziou, S. 1976b. Recherches sur les sources du cuivre dans l'IRAN ancien, IRAN XIV.

Berthoud, T. 1977. L'analyse multidimensionnelle appliquée aux données d'analyse chimique sur des objets archéologiques, Revue d'Archéométrie no. 1.

Berthoud, T., Besenval, R., Carbonnel, J.P., Cesbron, P. and Françaix, J. 1978a. Etude géochimique d'indices de cuivre d'Afghanistan. Implications structurales, C-R Acad. des Sc. PARIS t. 287, série D, pp. 187-190.

Berthoud, T., Besenval, R., Carbonnel, J.P., Cresbron, F. and Liszak-Hours, J. 1977. Les anciennes mines d'Afghanistan. Rapport préliminaire RCP 442 - Paris. Multicopie.

Berthoud, T., Besenval, R., Cleuziou, S. and Drin, N. 1978b. Les anciennes mines de cuivre du Sultanat d'OMAN. Rapport préliminaire RCP 442 - Paris Multicopie.

Berthoud, T. 1979. Etude par l'analyse et la modélisation de la filiation entre minerais de cuivre et objets archéologiques du Moyen Orient (IVè - IIIè millénaire). Thèse de doctorat d'Etat - Sciences Physiques - Université P. et M. Curie Paris.

102

Muhly, J.   1977.   The Copper Oxide Ingots and the Bronze Age
                                metals trade. Iraq 39 pp. 73-82.

Tylecote, R. F.   1977.   Summary of Experimental work on Early Copper
                                smelting in Aspects of Early Metallurgy ed.
                                W. A. Oddy pp. 5-12 London.

EARLY COPPER SMELTING TECHNIQUES IN SINAI AND IN THE NEGEV AS DEDUCED FROM SLAG INVESTIGATIONS

HANS-GERT BACHMANN

Wildaustrasse 5, D-6450 Hanau 9

Summary

During extensive surveys the "Arabah Expedition", founded and directed by Beno Rothenberg, has discovered numerous copper smelting sites in the Sinai peninsula and in the Timna region. Despite the fact that so far only a few of them have been excavated, we are nevertheless able to understand the basic principles of copper smelting technologies practised in these areas from the 4th millenium until Roman times. This was made possible by thorough slag identification and classification.

The following groups of slags could be distinguished:

- furnace conglomerate

- furnace slags, spinel type

- tapped slags, fayalite type

- tapped slags, knebelite type

- tapped slags, pyroxene type

- crucible slags.

These groups are described in detail (chemical analyses, phase compositions etc.). In appendix 1 the assignment of ternary phase diagrams to analytical data of silicate slags is treated in a more general way, together with full details of a computer program to carry out the necessary calculations. Appendix 2 is devoted to the derivation of an equation to relate the viscosity of silicate systems to temperature and slag composition in order to help understand progress in process development and optimisation.

Keywords: SINAI, TIMNA, COPPER SMELTING, SLAGS, ARCHAEO-METALLURGY, CHEMICAL ANALYSIS, PHASE DETERMINATION, PHASE DIAGRAMS, VISCOSITY, PROCESS OPTIMISATION

INTRODUCTION

The metal most influential to invention and development of artefacts in early cultures was copper. Its value as a versatile material for craft and trade initiated the search for sources of this metal. Rapidly increasing demand stimulated mining and metallurgy. Once the basic principles of smelting were understood, endeavours were made to increase production, output and efficiency. Economic laws evidently played the same role in ancient metallurgy as they do in modern process technology.

The archaeologist in search of early production centres has very few material evidences to aid him in his research. If he is fortunate, workshops for the manufacture of metallic articles may be discovered during excavations of settlements. Rarely, however, are the metals used for melting and/or alloying produced at habitation sites. One must look elsewhere for sites where metals were actually smelted. Activity at those sites prevailed only as long as resources such as ore, fuel and labour were available. If any of these were exhausted, the sites were abandoned, leaving behind few traces. Fortunately, every smelting process produced waste. This left-behind material was useless even to later inhabitants. It remained were it was deposited.

The waste material indicative of by-gone metallurgical activities is slag. A task for the archaeologist engaged in archaeo-metallurgy is the discovery of slag accumulations during his surveys. He can attempt to date such sites by pottery, lithic remains or by other evidences (stratigraphy etc.), but he will have to call on his partners - the metallurgist and chemist - to tell him what the slags reveal; after all, slags are indicators of the process or processes once practiced at the site in question. No matter what other relics can be salvaged from archaeo-metallurgical sites, no interpretation of former activities is complete without thorough characterisation of the slags found.

The Arabah Expedition - founded and directed by Beno Rothenberg - has always rigorously stuck to the policy that archaeo-metallurgy is a joint venture in which archaeologists, metallurgists, and chemists are equal partners. This paper deals with the technical aspects of early copper smelting techniques in the Sinai peninsula and the Negev desert. They have to be augmented by other criteria, chronology, cultural significance and other essential relations. The technology must be viewed in a proper framework, as we are - quoting Mortimer Wheeler - not digging-up things but man.

FIELD IDENTIFICATION OF SLAGS

Extractive metallurgy is the reduction of ores to metal. Ores are invariably mixed with gangue (i.e. impurities); these have to be separated from the metal during or after reduction of the ore. This process is termed slagging. Slag formation is essential to metal production and an understanding of the rules and laws governing slag formation are paramount for an evaluation of the efficiency of pyrometallurgical operations.

Table 1:   Field Identification of Slags

- Types (s)

- Amount(s)

- Average Size of Slag Fragments

- Average Size of Slag "Cakes"
         (Weight, Thickness Etc).

- Colour

- Streak

- Texture

- Porosity

- Inclusions

- Weathering

- Process Relation (Tapping Etc.)

Slag identification begins in the field. By ocular observation assisted by simple aids, such as a magnifying lens, a geologist's hammer, a measuring tape and perhaps a streak plate, a slag heap and its contents, i.e. the individual slag pieces it is composed of, can be characterised (cf. table 1; a more detailed treatment of this subject can be found elsewhere (Bachmann, 1980).

LABORATORY IDENTIFICATION OF SLAGS

The various approaches to the laboratory identification of slags available to investigators are summarised in table 2. A description of the majority of the methods mentioned can be found in Tite (1972). Some of the methods overlap with regard to the results and information they can give, others are unique and have no alternative. Not all the properties that can be determined by the methods outlined are of equal importance. However, if possible the chemical composition of a slag should be supported by determination of the phase composition, i.e. the identification of the dominating slag minerals present. Microscopic investigation of polished or thin sections gives detailed information about small fragments of the sample, while the overall phase composition is adequately revealed by X-ray powder diffractometry. The physical properties of slags may yield valuable additional information. This is especially true of melting points and viscosities. However, indications of these can be obtained already from the chemical and phase composition, so omitting their experimental determination is no serious negligence.

Table 2:   Laboratory Identification of Slags

CHEMICAL COMPOSITION

- Wet Chemical Analysis

- Spectroscopy (Emission, Atomic Absorption, Induced
   Coupled Plasma Etc.)

- Neutron Activation Analysis

- X-Ray Fluorescence Analysis (Wave-length & Energy-
   Dispersive Methods).

- Mass Spectrometry (For Isotope Analysis)

PHASE ANALYSIS

- Visual-light Microscopy (Thin and/or Polished Sections)

- Scanning Electron Microscopy

- X-Ray Diffraction (Film and Diffractometer Methods)

PHYSICAL CONSTANTS

- Specific Gravity

- Melting Behaviour (DTA and/or Heating Stage)

- Viscosity

AGE DETERMINATIONS

- Radiocarbon Dating (Charcoal, Carbides)

- TL-Dating (of Quartz Inclusions Etc.)

- Fission Track Dating (of Glasses)

Radiocarbon dating of charcoal remains occurring in or with slags follows the same routine as that of other carbon containing relics, but a wider margin of error has generally to be taken into account, because charcoal was not only made from freshly cut trees, but also from trees long dead. Recently, successful attempts have been made to apply thermoluminescence dating techniques to slags (Carriveau, 1974 and 1978). Slags occasionally tend to solidify as amorphous glasses instead of as crystalline compounds. Glass formation can be due to rapid cooling and/or chemical composition. Such slags may be suitable for fisson track dating, provided their uranium content is high enough to give rise to a sufficiently large number of tracks to be recorded (Wagner, 1979).

COPPER ORES IN SINAI AND NEGEV

We have applied these general principles of characterisation to slags from ancient metallurgical centres in the Negev and the Sinai peninsula. Fortunately, in these areas mining and smelting of copper ores were limited to secondary copper ores. We did not encounter the problems connected with sulphidic ores, such as roasting, matte smelting, various steps of refining. As evident from table 3a small amounts of copper sulphides occasionally do occur together with secondary ores in the said areas, but they are minor admixtures not requiring separate treatment. Table 3b lists the gangues and associated minerals that have to be considered in process reconstruction. Among these iron and manganese oxides are important as fluxes.

Table 3a : Copper Ores in Sinai and Negev

- Chrysocolla	$CuSiO_3 . n H_2O$
- Malachite	$Cu_2 (OH)_2 CO_3$
- Paratacamite	$Cu_2 (OH)_3 Cl$
- Atacamite	$Cu_2 (OH)_3 Cl$
- Brochantite	$Cu_4 (OH)_6 SO_4$
- Sulphides	e.g. Cu_2S
- Tenorite	CuO
- Cuprite	Cu_2O
- Native Copper	Cu
- Turquoise	$(Cu, Zn)(Al, Fe)_6 (OH)_2 (PO_4)_4 . 4 H_2O$

Table 3b: Associated Minerals (Gangue Etc.) in Sinai and Negev

- Quartz	SiO_2
- Various Silicates	-
- Calcite	$CaCO_3$
- Fluorite	CaF_2
- Gypsum	$CaSO_4 . 2 H_2O$
- Baryte	$BaSO_4$
- Halite	$NaCl$
- Hematite	$\alpha - Fe_2O_3$
- Pyrolusite	MnO_2

TYPES OF SLAGS

The slags collected from numerous sites and analysed according to the guide-lines given can be fitted into six separate groups:

- furnace conglomerate

- furnace slag, spinel type

- tapped slag, fayalite type

- tapped slag, knebelite type

- tapped slag, pyroxene type

- crucible slag, spinel-delafossite type

The "fingerprint" of slags belonging to the various groups outlined are given in the following chapters, together with tables of analytical data. To the description of the different types of slag is added the probable process of formation plus a brief evaluation of its significance with regard to efficiency and optimisation.

This schematic way of presentation was chosen to bring out the steps of process development more clearly. As our work on the large number of smelting sites in these areas is still under way, this classification simultaneously serves as a working hypothesis. With new results coming up, the categories will eventually have to be verified and expanded.

So far, a total of 125 slag samples from Sinai and Negev have been analysed chemically by wet-chemical methods and quantitative X-ray fluorescence (XRF); 70 of these were submitted to phase determination by X-ray powder diffraction (XRD). The wet-chemical analyses were carried out by A. Lupu and his coworkers, XRF and XRD by the author and his team. These investigations are summarised on tables 4 to 8. Unless otherwise stated, the results have not been published previously. In order to simplify the description of phases identified in the slags by XRD, the following abbreviations were introduced:

CAL = calcite, $CaCO_3$

CF = calcium ferrite, $CaFe_4O_7$

CR = cristobalite, high-SiO_2

CU = metallic copper, Cu

CUP = cuprite, Cu_2O

DEL = delafossite, $CuFeO_2$

FAY = fayalite, Fe_2SiO_4

HED = hedenbergite, $CaFeSi_2O_6$ (pyroxene)

HEM = hematite, α-Fe_2O_3

JOH = johannsenite, $CaMnSi_2O_6$ (pyroxene)

KIR = kirschsteinite, $CaFeSiO_4$

KNE = knebelite, $(Mn,Fe)_2SiO_4$

LAR = larnite, β-Ca_2SiO_4

MAG = magnetite, Fe_3O_4 (spinel)

PAR = paratacamite, $Cu_2(OH)_3Cl$

Q = quartz, α-SiO_2

SHA = shannonite, γ-Ca_2SiO_4

SIL = unidentified silicates

SOD = sodium chloride, NaCl

SP = spinels of varying composition, generally belonging to the group $(Fe,Mn,Mg)(Fe,Al,Mn)_2O_4$

$$TEP = \text{tephroite, } Mn_2SiO_4$$
$$WU = \text{wüstite, "FeO"}$$

This code has been used throughout tables 4 to 8.

FURNACE CONGLOMERATE

Description :   Partly decomposed ore and gangue particles, charcoal fragments, copper prills; slag formation insignificant.

Process :   Bowl hearth (blow-pipe operation?) or small furnace above ground (natural draught, wind channel?); temperature near or below 800°C, no equilibrium.

Summary :   Limited recovery of metal, poor separation, no segregation.

The term "furnace conglomerate" was coined to distinguish this material from slags in the true sense, i.e. molten mixtures of silicates (with or without oxides etc.). The "conglomerate" represents the very first step of the pyrotechnical treatment of ores which, however, has already resulted in a partial reduction of ores to metal. The type specimen of this group comes from Wadi Amram, Site 33, Timna region (Negev). It consists of a block of cinder-like material, approximately 25cm in diameter and about 15cm high. This lump of an extinguished bowl-furnace charge was cut into slices about 1cm thick in order to get an impression of it's components. The individual constituents (decomposed ore nodules, fragments of gangue, pieces of charcoal and globules of metallic copper, varying in size from below one to above five millimeters) are slightly fused and held together by a "slaggy" bonding agent. Whether this specimen represents an interrupted smelting attempt or the final product is insignificant for our aim to trace stages of technological development. No doubt, this find has preserved the very first step of conversion from ore to metal. Chronologically, this could have happened any time during the transition period from Stone Age to Copper or Bronze Age, but we may expect the same result if during a much later period men not yet skilled in the art of metal-making had experimented with green-coloured "stones" (i.e. secondary copper ores). As the genesis and composition of this material were obvious from it's appearance, it was not analysed further.

FURNACE SLAG, SPINEL-TYPE

Description :   Genuine slag with occasional relics of original charge (e.g. quartz fragments),
oxides of spinel type, e.g. Fe_3O_4, $(Fe, Mn)(Fe, Al)_2O_4$, fayalite, Fe_2SiO_4, pyroxenes, e.g. $FeCaSi_2O_6$, copper prills embedded in slag.

Process :   Small to medium-sized furnaces, (20 to 30cm diameter) forced draught (one or more tuyeres), no tapping worth mentioning, self-fluxing ores or use of ill-defined fluxes, temperature near or above 1200°C.

Summary :   High loss of copper in slag, no tapping, no segregation, crushing of slags necessary to recover metal.

A large number of slags belonging to this group from numerous sites in the Sinai peninsula and the Timna region have been investigated. The wide variation of the major constituents bears witness to a "trial-and-error" approach in copper smelting technique. Though slagging was achieved, the copper metal reduced from the ore was mainly embedded in the slag. This is indicated by the often very high copper contents found. As there was essentially no tapping involved, all these slags have

Table 4: Analytical Data for Furnace Slags

Area	Site	Sample	Cu	SiO$_2$	FeO	MnO	CaO	MgO	Al$_2$O$_3$	K$_2$O	Na$_2$O	TiO$_2$	P$_2$O$_5$	S	Zn	Pb	Reference	Phase analysis/Remarks
Sinai	350D	439	2.87	15.95	62.50	.45	5.74	.40	4.46	.08	4.18	-	.17	n.a.	-	-		SP
	418	530A	4.90	29.60	33.45	9.77	11.20	1.12	3.37	.60	.94	n.a.	.82	n.a.	.47	.15		FAY, SP
	418	530B	8.50	40.50	24.72	4.95	11.20	1.36	2.44	.78	2.60	n.a.	.74	n.a.	.30	.12		FAY, SP
	418	530C	4.40	34.30	34.50	9.93	10.08	1.08	1.95	.84	.67	n.a.	.81	n.a.	.20	.16		
	453	421	3.72	37.56	46.37	.50	2.66	.30	5.06	.18	.95	n.a.	.22	n.a.	<.1	-		
	454	521A	6.00	31.85	48.17	.30	6.10	1.00	1.71	.74	.63	n.a.	1.20	n.a.	.10	.01		
	454	521B	4.07	33.62	48.07	.29	4.57	.81	2.28	.74	.65	n.a.	1.22	n.a.	.1	.01		
	526	246	.80	38.93	33.01	.30	15.96	1.80	4.34	.09	n.a.	n.a.	1.13	n.a.	n.a.	n.a.		
	590	432	.47	34.33	53.16	.06	4.76	.10	3.18	.09	1.61	n.a.	-	n.a.	<.1	<.1		FAY, SP
	590	434	.75	35.08	52.07	.04	4.17	.48	3.79	.09	2.13	n.a.	.22	n.a.	.01	.02		FAY, MAG
	590	513F	2.58	33.11	47.29	.06	6.42	.98	4.21	.77	2.47	n.a.	.49	n.a.	.02	.02		FAY, SP, HED?
	590	514/1	1.42	35.05	54.25	.06	5.57	.34	2.58	.78	1.55	n.a.	.42	n.a.	.02	.03		FAY, SP
	590	514/2	1.49	35.19	51.48	.10	4.80	.58	1.80	.30	2.70	n.a.	1.10	n.a.	.06	.06		
	590	514/3	.95	35.21	51.00	.15	5.25	.40	3.17	.30	1.55	n.a.	.85	n.a.	.04	.06		
	590	514B	1.75	27.24	50.50	.03	8.71	.58	3.17	1.22	.67	n.a.	.62	n.a.	.06	.02		slagged lining
	599	442	.31	27.79	54.87	-	8.47	.31	3.00	.04	3.54	.4	.25	.3	-	<.1		FAY, SP
Timna	30A	27	2.7	35.1	52.2	1.1	5.0	-	1.9	-	n.a.	.4	1.7	.3	.2	.1	Lupu (1970)	FAY, SP
	39	09	1.10	44.78	22.38	n.a.	11.42	2.65	3.93	n.a.	n.a.	n.a.	n.a.	n.a.	~.1	~.1	"	FAY, SP, Q
	39	09b	4.03	23.21	33.17	n.a.	19.74	n.a.	9.06	n.a.	n.a.	n.a.	n.a.	n.a.	~.1	~.1	"	FAY, SP, Q
	39	14b	2.80	32.73	43.13	2.23	1.63	1.86	1.63	n.a.	n.a.	n.a.	n.a.	n.a.	~.1	~.1	"	FAY, SP, Q
	39	14x	4.21	28.90	48.75	.40	6.32	.29	4.77	n.a.	n.a.	n.a.	n.a.	n.a.	~.1	~.1	"	FAY, SP, Q
	39	50	1.31	30.71	49.51	2.00	4.06	n.a.	6.83	n.a.	n.a.	n.a.	n.a.	.22	~.1	~.1	"	DEL, FAY, Q, SP
	39	51	2.48	39.22	37.31	.32	6.39	n.a.	11.11	n.a.	n.a.	n.a.	n.a.	.32	~.1	~.1	"	slagged lining
	39	51b	.02	28.10	20.99	.10	23.66	1.77	21.00	n.a.	n.a.	n.a.	n.a.	n.a.	~.1	~.1	"	slagged lining
	39	52	5.04	48.42	22.84	.27	13.60	2.56	1.43	n.a.	n.a.	n.a.	n.a.	n.a.	n.a.	n.a.	"	
	39	53	5.12	33.21	41.80	1.40	9.80	.17	1.70	n.a.	n.a.	n.a.	n.a.	n.a.	n.a.	n.a.	"	
	39	54	2.36	48.25	35.50	.20	8.95	.81	2.12	n.a.	n.a.	n.a.	n.a.	n.a.	n.a.	n.a.	"	
	39	58a	15.12	34.25	21.69	n.a.	22.13	1.62	2.12	n.a.	n.a.	n.a.	n.a.	n.a.	n.a.	n.a.	"	slagged lining
	39	58b	16.60	45.92	14.20	n.a.	8.80	2.22	17.13	n.a.	n.a.	n.a.	n.a.	n.a.	n.a.	n.a.	"	slagged lining
	39	59	1.89	21.50	43.00	n.a.	12.04	1.37	9.06	n.a.	n.a.	n.a.	n.a.	n.a.	n.a.	n.a.	"	
	39	60	1.72	30.10	32.71	.82	9.00	.88	9.00	n.a.	n.a.	n.a.	n.a.	n.a.	~.1	~.1	"	SP, SIL; slagged lining?
	39	64	8.36	16.26	43.12	.16	20.80	.23	.37	n.a.	n.a.	n.a.	n.a.	n.a.	~.1	~.1	"	SP, SIL; slagged lining?
	39	69	6.81	40.96	33.06	n.a.	11.24	1.07	2.96	n.a.	n.a.	n.a.	n.a.	n.a.	n.a.	n.a.	"	
	39	72a	2.64	36.77	34.91	.80	12.10	1.42	.98	n.a.	n.a.	n.a.	n.a.	n.a.	n.a.	n.a.	"	
	39	72b	1.71	38.65	34.56	1.00	12.76	.18	.91	n.a.	n.a.	n.a.	n.a.	n.a.	n.a.	n.a.	"	
	39	72c	1.32	40.50	31.48	1.13	16.85	4.48	.41	n.a.	n.a.	n.a.	n.a.	n.a.	n.a.	n.a.	"	
	39	217	5.89	42.36	31.19	n.a.	14.48	n.a.	2.65	n.a.	n.a.	n.a.	n.a.	n.a.	~.1	~.1	"	Q, SP
	39	218	3.50	51.28	20.36	n.a.	4.21	n.a.	5.31	n.a.	n.a.	n.a.	n.a.	n.a.	~.1	~.1	"	Q, SP
Arabah	189A	1	3.76	46.03	13.70	21.20	4.21	.45	2.20	n.a.	n.a.	n.a.	n.a.	n.a.	n.a.	n.a.		Q, DEL, SP
	189A	2	6.46	41.91	12.69	27.68	3.17	-	.93	n.a.	n.a.	n.a.	n.a.	n.a.	n.a.	n.a.		JOH, HED
	189A	3	4.24	48.26	16.59	19.74	3.10	.37	1.83	n.a.	n.a.	n.a.	n.a.	n.a.	n.a.	n.a.		
	189A	4	8.28	49.41	16.63	20.97	2.02	-	2.10	n.a.	n.a.	n.a.	n.a.	n.a.	n.a.	n.a.		
	189A	5	9.20	51.60	25.64	2.02	3.15	.38	1.05	n.a.	n.a.	n.a.	n.a.	n.a.	n.a.	n.a.		Q, CR, MAG
	201	1126	2.2	34.5	45.8	.3	6.5	-	3.1	-	n.a.	.2	.4	.2	-	-		CF, KIR
	201	1128	5.3	34.7	36.2	.2	15.4	-	2.5	-	n.a.	-	.6	.2	-	-		LAR, SHA, CF
	201	1127	29.4	34.3	17.3	.2	9.5	-	.6	-	n.a.	.1	.4	.2	.4	.3		KIR, CF, CU, CUP

to be classified as furnace slags. Equilibrium has not been reached, therefore, no attempt was made to assign these slags to ternary phase diagrams. Chronologically these materials belong - at least in the areas concerned - to the Chalcolithic period.

The results are summarised on table 4. For a detailed description of the well investigated Timna Site 39 see Rothenberg 1972 and 1978. Chemical analyses of slags from this site have previously been published by Lupu 1970.

TAPPED SLAG, FAYALITE-TYPE

Description : Free-flowing tap slags, sometimes together with viscous furnace slags, mainly fayalite plus minor admixtures of oxides.

Process : Medium-sized furnaces, (about 20cm diameter) forced draught (tuyeres), tapping through tap hole, but also slag recovery from inside of furnace,
use of iron-oxide flux with silica-rich ores and silica flux with iron-rich ores,
temperature above $1250^{\circ}C$.

Summary : Varying degree of efficiency, depending on mode of operation and composition of charge, probably significant segregation.

In this group we are dealing with a very common and widely distributed assemblage of ancient metallurgical slags, characterised by the dominating presence of fayalite. The process was often mastered to such a degree that equilibrium and a high degree of reproducibility was maintained. Advanced technology resulting in separation of metal from slag and tapping of the latter enables us to treat these slags as relatively homogeneous products. Therefore, it is justified to apply the phase rule and place these slags into ternary diagrams. The procedure employed to correlate chemical composition to phases is outlined in appendix 1. By introducing the concept of "selection quotients" (cf. appendix 1) to each slag analysed, an appropriate ternary diagram is ascribed. Of the five ternary diagrams offered for selection, nearly all the slags belonging to this group can be fitted into the ternary diagram: Anorthite $(CaAl_2Si_2O_8)$ - SiO_2 - FeO. Though this treatment involves some simplifications, it is basically correct. The phases present as predicted by calculation could be confirmed by phase analyses using X-ray diffraction. Table 5 summarises the results obtained. In addition to the ternary systems characteristic for each slag (plus the coordinates in terms of percentages FeO, SiO_2 and anorthite, adding up to one hundred molar percent), two additional values are given: the basicity number of the slag and its viscosity coefficient. Their significance is explained in appendix 1. Figure 1 illustrates the position of slags of this group within the ternary diagram: $CaAl_2Si_2O_8$ - SiO_2 - FeO.

Though dating of smelting sites in Sinai and the Negev is often difficult due to lack of chronological evidences (pottery etc.), archaeologists are inclined to place the technology represented by the slags of this group into the 14th to 11th century B.C. Among the sites mentioned in table 5, Timna Site 2 has been dealt with extensively by Lupu & Rothenberg, (1970), and Rothenberg, (1972).

TAPPED SLAG, KNEBELITE-TYPE

Description : Large cakes of homogeneous tap slag, mainly of knebelite, $(Fe, Mn)_2 SiO_4$, and tephroite, Mn_2SiO_4.

Process : Large-sized furnaces, (40 to 60cm diameter) forced draught (several tuyeres), tap hole, heated slag pit,
manganese-oxide flux, use of return slags as additional flux?,
temperature above $1300^{\circ}C$, equilibrium,
temperature-composition-viscosity optimum.

Summary : Highly efficient, perfect separation and good reproducibility.

Table 5: Analytical Data for Fayalite Slags

Area	Site	Sample	Cu	SiO₂	FeO	MnO	CaO	MgO	Al₂O₃	K₂O	Na₂O	TiO₂	P₂O₅	S	Zn	Pb	Reference	AN	SiO₂	FeO	Bas.No.	Vis.Coef.	Phase analysis/Remarks
Sinai	317	245	6.20	38.99	31.96	.07	5.60	13.00	3.20	n.a.	n.a.	n.a.	n.a.	n.a.	n.a.	n.a.	n.a.	2	18	48	34	1.276	1.200	----
	317	246	.80	38.83	33.01	.30	15.94	1.80	4.34	n.a.	n.a.	n.a.	n.a.	n.a.	n.a.	n.a.	n.a.	2	42	21	37	1.150	1.183	----
	324	413	2.25	31.45	40.96	.17	4.80	.30	5.22	.09	1.37	~.1	.17	n.a.	<1.0	-	n.a.	2	25	26	49	1.198	1.301	FAY, HED
	385	119/1	1.16	35.48	33.51	1.09	6.66	1.01	6.61	.63	1.02	~.1	1.07	n.a.	-	-	n.a.	2	28	24	48	1.205	1.281	KNE
	385	119/2	1.66	38.36	36.21	.37	11.57	1.71	4.18	.36	.84	n.a.	2.15	n.a.	n.a.	n.a.	n.a.	2	35	24	41	1.141	1.200	
	458	423	.27	36.22	39.31	5.17	4.37	1.79	7.20	.15	1.95	-	.10	n.a.	<.1	-	n.a.	2	26	26	48	1.151	1.215	FAY
	485	253/1	4.82	36.57	36.78	.05	7.30	.61	3.07	n.a.	n.a.	n.a.	1.84	n.a.	n.a.	n.a.	n.a.	2	24	32	44	1.030	1.129	
	485	253/2	2.40	45.63	28.01	.11	9.80	.40	3.74	n.a.	n.a.	n.a.	1.13	n.a.	n.a.	n.a.	n.a.	2	30	38	32	.723	.776	
	485	424/S	1.96	44.51	37.34	.08	5.07	.78	6.24	.10	1.97	~.1	.27	n.a.	-	-	n.a.	2	25	35	40	.827	.893	FAY, Q
Timna	2	LR3	.27	29.70	53.10	2.50	4.70	2.30	9.80	n.a.	n.a.	n.a.	n.a.	.20	n.a.	n.a.	n.a.	Lupu & Rothenberg (1970)	25	18	57	1.549	1.585	
	2	LR4	.61	40.20	43.30	1.70	9.30	.50	2.20	n.a.	n.a.	n.a.	n.a.	.10	n.a.	n.a.	n.a.	"	23	30	47	1.165	1.292	
	2	LR5	2.30	35.70	32.90	2.70	13.70	.30	3.10	n.a.	n.a.	n.a.	n.a.	.69	n.a.	n.a.	n.a.	"	38	22	40	1.197	1.278	
	2	LR6	3.80	32.50	46.20	.40	2.30	4.70	2.90	n.a.	n.a.	n.a.	n.a.	.14	n.a.	n.a.	n.a.	"	10	32	58	1.416	1.514	
	2	LR7	.32	36.10	40.40	3.50	4.50	3.10	2.30	n.a.	n.a.	n.a.	n.a.	.25	n.a.	n.a.	n.a.	"	15	33	52	1.233	1.341	
	2	LR8	.28	34.20	36.50	7.60	5.20	1.60	9.60	n.a.	n.a.	n.a.	n.a.	.17	n.a.	n.a.	n.a.	"	28	24	48	1.127	1.162	
	2	LR9	.85	35.40	52.40	.70	4.90	1.40	4.10	n.a.	n.a.	n.a.	n.a.	.30	n.a.	n.a.	n.a.		17	28	55	1.368	1.504	
	30	A	1.4	36.5	43.2	9.2	5.3	-	2.7	n.a.	n.a.	1.0	.5	-	.1	.1		2	16	30	54	1.302	1.472	FAY
	30	B	2.4	40.4	39.1	9.2	5.1	-	2.4	n.a.	n.a.	.8	.3	-	.1	.1		2	15	35	50	1.099	1.248	FAY, MAG?
	30	C	1.5	39.9	40.8	10.3	4.0	-	1.9	n.a.	n.a.	.8	.3	-	.1	.1		2	12	35	53	1.148	1.318	FAY, Q?
	30	D	.8	38.0	42.9	8.2	5.4	-	2.4	n.a.	n.a.	.8	.7	-	.1	.1		2	15	32	53	1.233	1.399	FAY
	30	E	2.1	32.7	45.8	7.4	3.9	-	2.2	n.a.	n.a.	.7	.5	-	.2	.1		2	16	30	54	1.302	1.484	FAY, MAG
	30	F	2.9	36.0	42.7	8.1	4.2	-	2.4	n.a.	n.a.	.8	.5	-	.2	.1		2	14	32	54	1.258	1.432	FAY, MAG, Q
	30	G	2.4	48.0	39.0	1.7	5.5	-	1.6	n.a.	n.a.	.3	-	-	.3	.1		2	15	43	42	.816	.931	Q, SP, FAY; slagged lining
	30	H	3.7	63.5	10.6	.8	5.9	-	10.1	.4	n.a.	.8	.1	.5	.1	.2		2	16	35	49	1.068	1.208	FAY
	30	I	.9	41.0	39.1	8.6	5.2	-	2.8	n.a.	n.a.	.7	.3	.1	.1	.1		2	16	35	49	1.235	1.406	FAY
	30	K	1.5	38.5	40.5	9.0	5.4	-	1.6	n.a.	n.a.	1.0	.2	.2	.1	.1		2	17	32	51	1.228	1.403	FAY
	30	L	1.7	37.5	44.4	7.1	4.2	-	1.6	n.a.	n.a.	.7	.4	-	.1	.1		2	13	33	54	1.167	1.323	FAY
	30	M	1.0	38.3	36.5	10.9	5.9	-	2.0	n.a.	n.a.	.8	.5	-	.1	.1		2	16	33	51	1.167	1.323	FAY
	30	10	9.6	41.7	32.7	10.9	10.7	-	1.0	-	n.a.	.3	-	-	.2	.3		2	28	34	38	.925	1.026	HED, CUP, FAY?

Table 6: Analytical Data for Knebelite Slags

Area	Site	Sample	Cu	SiO₂	FeO	MnO	CaO	MgO	Al₂O₃	K₂O	Na₂O	TiO₂	P₂O₅	S	Zn	Pb	Reference	Syst.	AN	SiO₂	FeO+MnO	Bas.No.	Vis.Coef.	Phase analysis/Remarks
Sinai	344	214	1.06	42.29	22.65	14.32	6.14	.60	5.29	.66	.53	<.5	.70	n.a.	<1.	<.1		2	25	34	41	.869	.944	
	345	416	.68	28.96	35.27	11.40	6.73	2.24	7.80	.12	3.24	<.5	—	n.a.	<.5	—		2	34	15	51	1.576	1.605	TEP, KNE, SP
	347	417	1.85	33.55	19.33	14.01	12.98	3.91	5.72	.13	2.07	<.5	—	n.a.	<.5	—		2	44	15	41	1.350	1.335	TEP, SP
	350	121/1	1.05	29.42	28.02	19.56	3.70	.31	6.57	.78	.23	—	2.51	n.a.	<1.	—		2	22	24	54	1.356	1.462	KNE
	350	121/2	.62	24.84	34.19	19.42	3.90	.20	7.09	.68	.38	.5	2.90	n.a.	<1.	—		2	24	17	59	1.733	1.841	KNE
	350	121/3	1.01	32.57	14.89	29.62	3.46	.49	6.53	.86	.87	.5	1.13	n.a.	<1.	—		2	23	27	50	1.190	1.284	KNE
	350	T3/L1	.4	35.8	28.3	24.5	4.1	2.0	5.5	.3	n.a.	.4	1.1	.2	.1	—		2	18	28	54	1.331	1.433	
	350	T3/L2	.4	39.2	17.6	25.0	6.5	2.7	5.1	.3	n.a.	.4	1.0	.3	.1	—		2	23	30	47	1.115	1.176	
	350	T3/L3	.5	37.5	34.6	11.3	5.8	1.1	6.2	.4	—	.3	1.1	.3	—	—		2	23	28	49	1.132	1.217	
	350	T3/L4	.5	28.9	24.3	30.3	5.3	2.0	6.3	.2	—	.4	1.2	.2	<.1	—		2	22	20	58	1.679	1.764	
	354	122	1.25	36.18	18.18	25.83	5.64	1.71	6.34	.67	.46	.5	1.21	n.a.	<.5	—		2	25	27	48	1.166	1.234	FAY, KNE, JOH
	354	418	1.95	37.70	10.77	27.50	5.04	1.36	7.65	.19	3.67	.5	.28	.20	<1.	—		2	32	26	42	1.028	1.070	TEP, KNE
Timna	2	LR1	.24	32.70	14.80	32.30	4.10	2.50	4.80	n.a.	n.a.	n.a.	n.a.	.20	n.a.	n.a.	Lupu & Rothenberg (1970)	2	18	28	54	1.346	1.432	
	2	LR2	.27	35.60	6.50	40.40	4.60	1.50	4.10	n.a.	n.a.	n.a.	n.a.	.13	n.a.	n.a.		2	17	31	52	1.241	1.350	
	30	2	.4	42.4	3.9	32.5	8.9	—	3.8	.6	n.a.	3.0	3.0	.6	—	—		2	27	33	40	.904	.983	Glassy slag
	30	20	4.0	40.7	9.0	33.3	6.7	—	2.5	.1	n.a.	1.5	2.0	.5	.1	.5		2	19	35	46	1.017	1.134	TEP, KNE
	30	25	.5	44.5	6.1	36.5	5.4	—	2.9	—	n.a.	1.7	—	.3	.1	.1		2	16	39	45	.904	1.013	TEP + glassy slag
	30	26/G1	n.a.	40.0	14.0	—	8.0	n.a.	9.0	n.a.	n.a.	n.a.	n.a.	n.a.	n.a.	n.a.								glassy slag;incompl.anal.
	30	26/Cr	.8	40.2	7.5	38.0	6.9	—	2.7	.1	n.a.	2.2	3.0	.5	.1	.4		2	19	33	48	1.097	1.221	TEP, KNE
	30	100	2.0	42.6	7.4	36.0	5.4	—	2.7	—	n.a.	1.6	—	.2	.1	.2		2	17	37	46	.961	1.077	TEP, KNE
Arabah	58	58C	.73	30.11	30.35	29.33	12.17	.98	n.a.	n.a.	n.a.	n.a.	n.a.	n.a.	n.a.	n.a.		2	24	17	59	2.148	2.418	

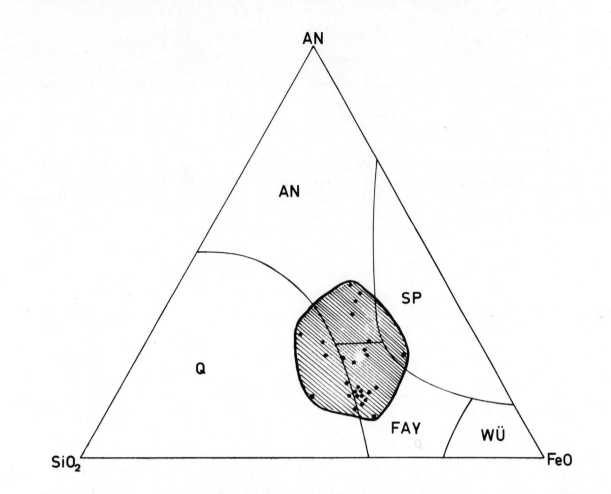

Figure 1: Position of tapping slags, fayalite type, in the ternary systems: Anorthite - SiO$_2$ - FeO.

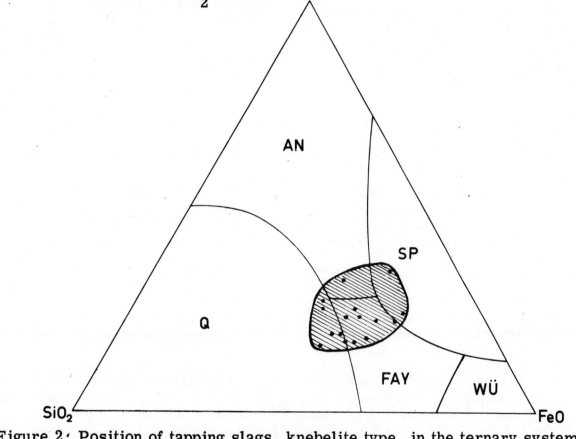

Figure 2: Position of tapping slags, knebelite type, in the ternary system: Anorthite - SiO$_2$ - FeO (+MnO).

114

While the slags of the groups described so far are likely to be found wherever copper smelting was carried out in the past, slags with an anomalous high manganese content are as yet specific to some archaeo-metallurgical sites in the Sinai peninsula and the Timna region. Both areas abound in deposits of rich manganese ores. The Sinai deposits were mined until 1967. In the Timna mountains exploration and trial exploitation went on until the early seventies. In ancient times these manganese oxides were employed as flux when smelting the siliceous secondary copper ores. Though iron oxides could have served this purpose just as well, copper smelters during the 10th to 8th century B.C. decided to use manganese fluxes instead. The reason for this was perhaps the easy access to large and uniform deposits of this mineral, mandatory for the reproducibility of the improved process practised during these centuries. A look at the chemical composition of fayalite slags (cf. table 5) also shows manganese contents occasionally approaching ten percent. Therefore, the use of manganese fluxes may already have started at an earlier date, but as the copper ores occurring in sedimentary rocks in Sinai and at Timna are often accompanied by manganese oxides, the manganese in the fayalite slags may also be due to the smelting of mixed, i.e. copper-manganese, ores.

In slag formation, manganese behaves very much like iron. It combines with silica to form manganese silicates isomorphous with the corresponding iron silicates. The Mn-equivalent to fayalite is tephroite, Mn_2SiO_4; if additional iron is present, the slag mineral knebelite, $(Fe, Mn)_2SiO_4$, with varying amounts of either Mn or Fe, is formed. Actually, the classification of smelting remains into fayalite and knebelite groups of slags is somewhat arbitrary. The introduction of three groups with low-, medium- and high-manganese contents would have been equally justified. We correlate high-manganese slags with an almost optimum smelting technique. High process efficiency could have been achieved in the same manner by using iron oxide fluxes, but for the reasons mentioned manganese ores were chosen. Some of the slags belonging to this group occur as amorphous glasses. This points to rapid cooling. Glassy slags often form the uppermost layer of huge slag cakes (weighing up to 25 kilograms and more). This apparently was the last material to be tapped when discharging a furnace. It could take up some constituents from the furnace lining (Al_2O_3 etc.) promoting glass formation. However, the composition of glassy slags varies little from that of associated crystalline slags (cf. table 6 and figure 2).

TAPPED SLAG, PYROXENE-TYPE

Description : Large cakes of homogeneous tap slag, mainly pyroxenes, $FeCaSi_2O_6$ etc. and fayalite.

Process : Large-sized furnaces, (40 to 60 diameter) forced draught (several tuyeres), tap hole, iron-oxides plus calcite (lime) used as fluxes, temperature above 1300^oC, equilibrium, temperature-composition-viscosity-correlation.

Summary : Very efficient, good to perfect separation, good reproducibility.

After many centuries of inactivity, during which all the previously attained "know how" was lost, a new endeavour at smelting Timna ores was made. This happened during the 2nd century A.D. when the country was under Roman rule. Though the same raw materials were still available in sufficient quantities, the use of manganese oxides as fluxes was not taken up again. The slags of this late phase are characterised by higher CaO-contents than any slags from earlier periods. This change in flux composition resulted in the formation of slags with pyroxenes as the main constituents. The use of lime as flux (in addition to iron oxide) was probably imported by foreign smelters starting a new enterprise at this traditional site. So far, this kind of process has only been observed at the site of Beer Ora, a few miles to the South of Timna. More about this interesting site can be found in Rothenberg, (1972). The analytical data are given in table 7 and figure 3.

Table 7: Analytical Data for Pyroxene Slags

Area	Site	Sample	Cu	SiO$_2$	FeO	MnO	CaO	MgO	Al$_2$O$_3$	K$_2$O	Na$_2$O	P$_2$O$_5$	Syst.	AN	SiO$_2$	FeO	WO	Bas.No.	Vis.Coef.	Phase analysis/Remarks
Beer Ora	28	127A	2.87	37.88	31.98	1.08	19.14	2.39	1.14	1.10	.78	1.18	3	37	–	37	26	1.379	1.447	HED, SP
	28	127B	.99	38.63	27.90	1.74	18.70	2.13	2.83	1.12	.84	1.93	3	40	–	26	34	1.229	1.265	HED, SP
	28	132	1.60	35.73	31.76	1.82	14.30	1.57	1.93	.96	.60	1.41	2	40	20	40	–	1.273	1.354	HED, SP
	28	133	1.90	28.89	38.04	1.55	14.69	.85	.60	.85	.60	1.74	3	32	–	47	21	1.752	1.919	HED, SP
	28	134	1.28	24.73	37.83	2.03	16.43	2.28	2.79	.84	.48	1.30	3	34	–	48	18	2,017	2.104	SP, HED?
	28	135	2.30	49.41	8.55	.42	14.86	3.59	8.73	.84	.26	.45								slagged furnace lining
	28	136	12.19	53.13	10.72	1.79	6.13	.99	1.84	.27	.13	.46								Q, CR, MAG; slagged lining
	28	137	5.06	38.59	21.47	1.01	18.88	2.98	3.87	.24	.15	.99	3	42	–	29	29	1.071	1.053	HED, SP

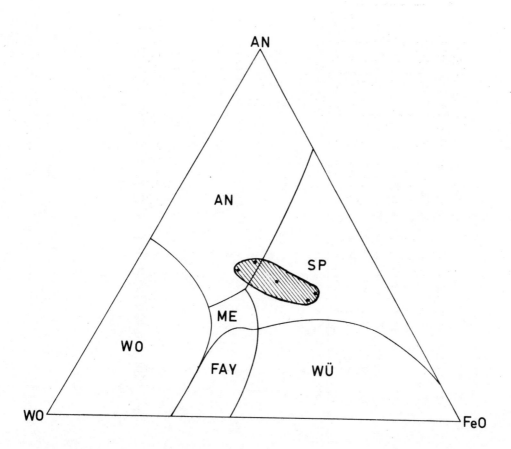

Figure 3: Position of tapping slags, pyroxene type, in the ternary system.: Wollastonite - Anorthite - FeO.

CRUCIBLE SLAG

Description : Small fragments, either isolated or adhering to crucible sherds, varying composition, hard to define, constituents mostly oxides (spinels, delafossite, $CuFeO_2$, tenorite, CuO, cuprite, Cu_2O, plus some fayalite etc.
high copper content (as metal and/or oxides).

Process : Metal purification (melting <u>not</u> smelting in crucibles, use of single(?) tuyere to heat crucible contents.

Summary : Crucible melting often accompanies furnace smelting on the same site; however, some sites only show crucible melting.

The last group in our survey of slag types is reserved for material labelled "crucible slags". This often ill-defined metallurgical waste has a wide variation in composition. Some specimens may resemble genuine smelting slags which perhaps very often they are, adhering to pieces of crude copper metal which was purified by crucible melting. As the term implies, these finds are in most cases accompanied by sherds of vessels in which melting of metals for refining or alloying purposes was carried out. The examples given on table 8 are from Egyptian miners' camps in the Sinai peninsula, dating back to the Old, Middle and New Kingdoms.

Table 8: Analytical Data for Crucible Slags

Area	Site	Sample	Cu	SiO$_2$	FeO	MnO	CaO	MgO	Al$_2$O$_3$	K$_2$O	Na$_2$O	P$_2$O$_5$	Zn	Pb	Phase analysis/Remarks
Sinai	349	201a	.37	53.93	10.43	.37	5.20	3.00	12.58	1.21	1.13	.67	n.a.	n.a.	Q, SP, CAL?
	349	201b	5.93	43.46	15.19	.30	12.11	3.36	8.48	1.15	1.09	.98	n.a.	n.a.	Q, SP, CAL?
	349	201c	7.36	48.70	7.48	.40	6.38	—	13.29	.88	1.02	.37	n.a.	n.a.	Q, SP, CAL
	349	201d	1.33	49.55	14.07	.46	5.75	.97	12.46	.36	.75	.87	n.a.	n.a.	Q, HEM, CAL
	349	201/1	.59	71.30	14.05	.52	3.01	.52	5.80	.51	.80	1.00	n.a.	n.a.	Q, SP, CAL
	349	201/2	2.68	43.31	25.57	.32	5.89	1.11	8.77	.48	1.73	1.62	n.a.	n.a.	Q, CAL, FAY
	349	201/3	.81	59.96	12.14	.59	4.03	.59	5.58	.32	.32	.71	n.a.	n.a.	Q, SP, SIL
	349	201/4	6.75	69.75	5.51	.37	5.58	.61	3.89	1.38	.28	.50	n.a.	n.a.	Q, CR
	349	201/5	2.30	36.96	9.64	7.16	3.20	.79	9.18	.59	1.04	.60	n.a.	n.a.	SP, WÜ
	349	209	.97	34.74	45.03	3.03	5.57	—	2.76	n.a.	n.a.	—	n.a.	n.a.	resembling fayalite slag
	349	440	8.04	28.54	27.48	7.13	9.74	1.61	12.20	2.05	1.38	—	n.a.	n.a.	MAG, CUP, Q
	349	526C	5.73	58.78	4.50	.09	11.20	1.10	13.91	1.44	.67	.42	.4	.02	—
	355	445	2.48	50.33	6.36	.16	13.00	1.77	18.08	1.38	1.16	.53	<.1	n.a.	—
	357	115/1	3.79	22.03	44.36	.54	8.64	.45	4.04	.83	2.16	3.52	<.1	—	Q, FAY?, SIL?
	357	115/2	.35	41.08	10.75	.26	8.52	8.68	13.07	.41	1.79	.85	n.a.	n.a.	—
	357	115/3	.24	43.29	11.47	.26	8.24	9.35	13.96	.40	1.70	.97	n.a.	n.a.	—
	357	115/4	10.20	40.80	5.00	1.11	11.14	3.02	9.55	.31	.22	.33	n.a.	n.a.	SP, PAR, CUP
	357	120/1	1.06	48.04	17.80	2.67	7.91	.89	7.26	.20	3.84	1.22	<.1	—	Q, HED
	357	120/2	4.48	62.55	6.62	.75	7.89	.60	5.31	.30	1.34	.44	<.1	—	Q, SIL
	357	419	6.66	47.34	10.26	1.78	8.71	.42	9.10	.13	13.23	.26	<.1	<.1	Q, CUP, SOD
	357	529C	5.10	42.79	13.51	3.38	7.35	.73	9.40	.74	7.25	.30	.12	.04	—

CONCLUSION

As was pointed out at the beginning, progress in archaeo-metallurgical techniques should be evident from slag characterisation. To optimise the formation of slags is, therefore, essential to increase efficiency in metallurgical processes. A slag of optimum composition is one which has a low energy of formation, a low melting point and a high degree of fluidity (i.e. low viscosity). Slag formation is not only governed by the composition of the smelting charge, but also by the construction of the furnace and its mode of operation; in short: heat and mass balance. These are criteria applying to metallurgical processes in general. As we are concerned with archaeo-metallurgy, a few restrictions have to be kept in mind:

- The temperature could not be raised above 1300 to 1400°C; the limiting factor influencing combustion being human labour when applying forced draught with the aid of bellows or the dependance on prevailing winds when resorting to induced, natural draught,

- the only available energy source (and reduction compound) was charcoal,

- of the many possible fluxes facilitating slag formation only iron and manganese oxides, lime, quartz and perhaps fluorspar were known to early smelters, and finally

- refractory materials necessary for furnace construction etc. were limited to locally available clays and tempers.

As is evident from the gradual development of techniques practised in the Sinai peninsula and in the Timna region (and very likely elsewhere) each step was a further approach to maximum metal output with minimum energy input. The influencing process control parameters are:

- composition and purity of ore

- composition and purity of fluxes

- ratio of charcoal to ore plus fluxes

- furnace construction and insulation

- oxygen (= air) control
  (type, number and diameter of tuyeres, number of bellows and bellow-operators)

- mode of operation during run-time of furnace (e.g. sequence of charging, retention time of liquid slag prior to tapping etc.)

The interdependence of these process control parameters is schematically represented in figure 4.

This study may give the impression that the importance of slags is overemphasized in judging ancient pyrotechnological processes. We are fully aware that other remains, like furnace bottoms, fragments of furnace lining, tuyeres etc. are at least equally valuable evidences to help us reconstruct early technologies. They are, however, often missing or they require extensive excavation. To overcome these difficulties we aimed at obtaining the maximum possible information from analysing archaeo-metallurgical waste. The final proof whether our assumptions are correct will be the experimental reproduction of the postulated smelting procedures under controlled conditions. Thus, the results of slag investigations will serve merely as a guideline to unravel our technological past, which in turn is but an integral part of men's social and political history.

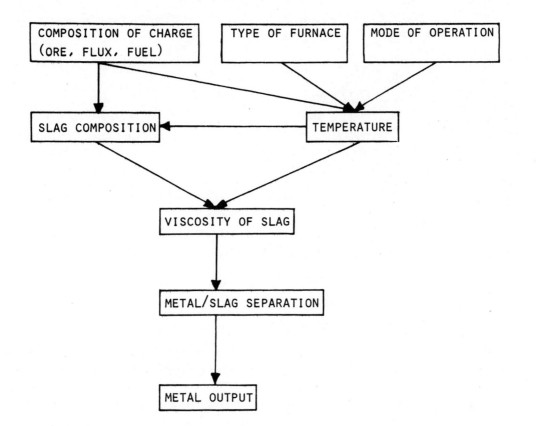

INTERDEPENDENCE OF PROCESS PARAMETERS

Figure 4: Interdependence of process parameters in early copper smelting.

APPENDIX 1

ASSIGNMENT OF TERNARY PHASE DIAGRAMS TO ANALYTICAL DATA OF SILICATE SLAGS

Co-author : Sandra Zacharias

To explain and understand the behaviour of ancient metallurgical slags during their solidification, the quaternary system: $CaO - FeO - Al_2O_3 - SiO_2$ was chosen. In order to include additional constituents present in most slags analysed, this system was expanded to: $CaO (+BaO+Na_2O+K_2O) - FeO (+MnO+MgO) - Al_2O_3 - SiO_2$. The graphical presentation of this system is a regular tetrahedron with the following four triangles (i.e. ternary systems) as faces:

$$CaO - Al_2O_3 - FeO$$
$$SiO_2 - Al_2O_3 - FeO$$
$$SiO_2 - Al_2O_3 - CaO$$
$$SiO_2 - CaO - FeO$$

Of these, the system $SiO_2 - CaO - FeO$ is of particular importance. It contains - among others - the slag compounds anorthite (AN), $CaAl_2Si_2O_8$, and gehlenite (GE), $Ca_2Al_2SiO_7$. If the mineral wollastonite (WO), $CaSiO_3$, is included, five ternary systems, all with FeO as one of their corners, are obtained. With their corners AN, GE, WO, FeO, SiO_2 and Al_2O_3 they are either external faces or internal planes of the quaternary main system $CaO - FeO - SiO_2 - Al_2O_3$:

(1) Al_2O_3 - AN - FeO

(2) AN - SiO_2 - FeO

(3) AN - WO - FeO

(4) GE - WO - FeO

(5) AN - GE - FeO

Most ancient slags can be accomodated in systems (2) and (3). System (1) requires Al_2O_3 - contents higher than normally found in slags; systems (4) and (5) are only valid for silicates with considerable CaO percentages. To find out which system can be ascribed to a particular slag, the chemical analysis is treated as follows:

- Summation of the oxide components, omitting P_2O_5 and TiO_2; conversion of Fe_2O_3 - if determined - to Fe_3O_4. The sum is set to one hundred percent, which implies multiplying the individual oxide components with an appropriate factor. This is called the "reduced analysis".

- Calculation of molar percentages of the oxides, using the data of the "reduced analysis"; from these, the following values are obtained: MA = molar percentage of Al_2O_3, MS = molar percentage of SiO_2 and MC = sum of molar percentages of CaO, BaO, Na_2O and K_2O. The molar fractions of MnO and MgO are added to the molar percentage of FeO.

- Calculation of "selection quotients" MA/MS, MA/MC and MC/MS. These "selection quotients" have specific values for the five systems mentioned. Each system is unambiguously characterised by a combination of "selection quotients" as shown in table 9.

- Calculation of the triangular coordinates AN, WO, GE, SiO_2, Al_2O_3 and FeO for the systems previously determined by the appropriate "selection quotients"; these add up to one hundred molar percent.

- The slag composition given by these coordinates can thus be presented graphically in the regular triangle of the system selected.

Table 9: Assignment of ternary phase diagrams with the aid of "selection quotients".

TERNARY PHASE DIAGRAM SELECTION QUOTIENT	(1)	(2)	(3)	(4)	(5)
MA / MS	$\frac{1}{2}$ — ∞	0 — $\frac{1}{2}$	0 — $\frac{1}{2}$	0 — 1	$\frac{1}{2}$ — 1
MA / MC	1 — ∞	0 — 1	0 — 1	0 — $\frac{1}{2}$	$\frac{1}{2}$ — 1
MC / MS	0 — $\frac{1}{2}$	0 — $\frac{1}{2}$	$\frac{1}{2}$ — 1	1 — 2	$\frac{1}{2}$ — 2

These calculations are best carried out by a computer program. The flow sheet of a **FORTRAN** program performing these calculations is given in figure 5, followed by a list of the program statements and a sample output. (For information about input variables and formats plus details of the calculation procedure see comments in program list.) An earlier version of the program was published by Bachmann (1978).

As modern slags are often specified by their basicity number or V-ratio (cf. Rosenqvist, 1974), calculation of this value is included in the program. The basicity number B of a slag is given by the equation:

$$B = \frac{\Sigma \; RO + R_2O}{\Sigma \; SiO_2 + Al_2O_3} \qquad \text{(in molar percentages)}$$

RO = FeO, MnO, CaO, MgO, BaO
R_2O = Na$_2$O, K$_2$O

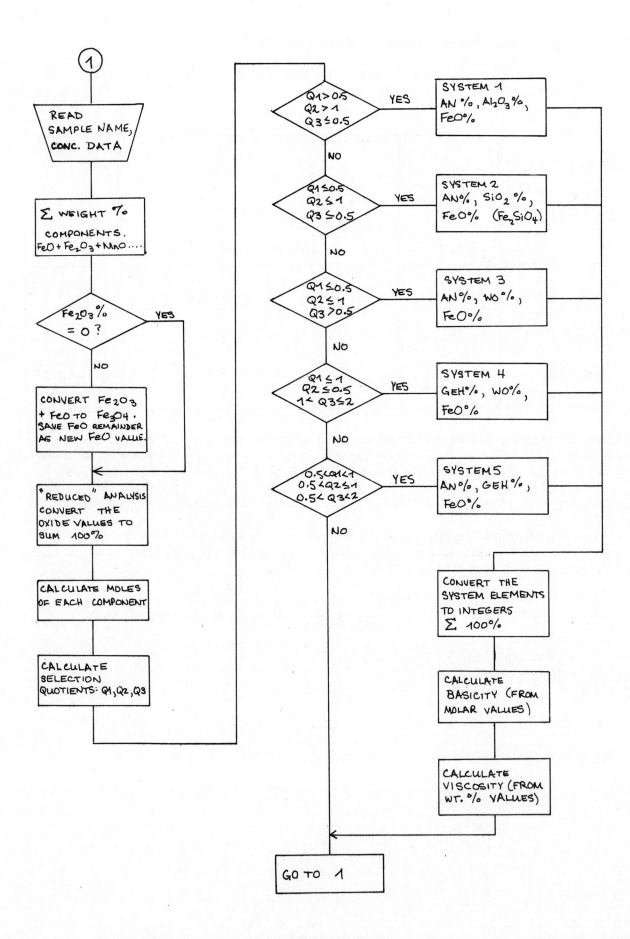

Figure 5: Flow sheet of computer program to derive ternary systems from slag analyses.

```
* * * * * * * * * * * * * * * * * * * * * * * * * * * *
*                                                     *
*            COMPUTER PROGRAM TO DERIVE               *
*                                                     *
*              TERNARY SYSTEMS                         *
*                                                     *
*             FROM SLAG ANALYSES                      *
*                                                     *
* * * * * * * * * * * * * * * * * * * * * * * * * * * *


      INTEGER     SYST, NAME
      REAL        MAL2O3, MSIO2, MCAO, MBAO, MK2O,
     *            MNA2O, MC, G1, G2, G3, MMGO, MMNO, MFEO
      DIMENSION AN(2), FEO(2), A(2, 16), NAME(10), C(12)
      CALL ASSIGN (5, 'SY: SILAN2. DDD; 1')
      OPEN (UNIT=6, NAME='SY: SILAN2. LST', TYPE='NEW', DISPOSE='PRINT')

      DO 1      I=1, 2
      DO 1      J=1, 16
      A(I, J)=0.
    1 CONTINUE


      THE PROGRAM WILL ACCEPT UP TO 16 ANALYTICALLY DETERMINED
      CONSTITUENTS.   THE FOLLOWING OXIDE COMPONENTS ARE PART
      OF THE PROGRAM:   FEO, FE2O3, MNO, MGO, AL2O3, SIO2, CAO,
      BAO, K2O, NA2O.
      UP TO 6 ADDITIONAL COMPONENTS ARE USER-ENTERED NAMETAGS
      IN FORMAT 2A4   E.  G.  :

      TIO2!!!!P2O5!!!!CU!!!!!!!!S!!!!!!!!!ZN!!!!!!PB!!!!!!

      THREE DATA CARDS ARE READ FOR EACH SAMPLE:
      1.    SAMPLE NAME IN FORMAT 10A4
      2.    ADDITIONAL COMPONENTS (UP TO 6) FROM ANALYSIS, IN FORMAT 2A4
      3.    CONCENTRATION OF EACH COMPONENT IN FORMAT F5. 2

      THE LAST DATA CARD READS 0000, IN THE FIRST FOUR COLUMNS


      READ (5, 1000) NAME
      IF (NAME(1) .EQ. '0000')  GO TO 200
    6 READ (5, 1010) (C(I), I=1, 12)
      READ (5, 1015) (A(1, J), J=1, 16)
      SUMMA=0
      DO 10  J=1, 16
      SUMMA=SUMMA+A(1, J)
   10 CONTINUE


      IF FE2O3 IS NOT EQUAL TO ZERO, CONVERT THE FE2O3 TO
      FEO AND FE3O4.
           MOLES FE2O3 = FMOL1
           MOLES FEO   = FMOL2
           REMAINDER FEO = REST
           %REMAINDER FEO =RTFEO
           FEO COMPONENT OF MAGNETITE = FEOMAG


      IF (A(1, 2) .EQ. 0. 0)            GO TO 15
      FMOL1=A(1, 2)/159. 694
      FMOL2=A(1, 1)/71. 847
      REST=FMOL2-FMOL1
      RTFEO=(A(1, 1)*REST)/FMOL2
      FEOMAG=A(1, 1)-RTFEO
      FE3O4=A(1, 2)+FEOMAG
      A(1, 1)=RTFEO


      IF AL2O3 EQUALS ZERO OR WAS NOT DETERMINED, SET ITS VALUE
      TO 0. 01.


   15 IF (A(1, 5) .EQ. 0. 0)      A(1, 5)=0. 01


      SUM THE TEN MAJOR OXIDE COMPONENTS
```

```
      TSUMMA=0
      TSUMMA=TSUMMA+A(1,1)
      DO 20   J=3,10
      TSUMMA=TSUMMA+A(1,J)
20    CONTINUE


C     CONVERT THE VALUES OF THE TEN MAJOR OXIDE COMPONENTS TO SUM 100%.


      F=100/TSUMMA
      A(2,1)=A(1,1)*F
      DO 30   J=3,10
      A(2,J)=A(1,J)*F
30    CONTINUE
      SUMMB=0
      SUMMB=SUMMB+A(2,1)
      DO 40   J=3,10
      SUMMB=SUMMB +A(2,J)
40    CONTINUE


C     CALCULATE MOLES FOR EACH OXIDE


      MFEO  =A(2,1)/71.85
      MMNO  =A(2,3)/70.94
      MMGO  =A(2,4)/40.30
      MAL2O3=A(2,5)/101.94
      MSIO2 =A(2,6)/60.06
      MCAO  =A(2,7)/56.08
      MBAO  =A(2,8)/153.36
      MK2O  =A(2,9)/94.19
      MNA2O =A(2,10)/61.99
      MC = MCAO+MBAO+MK2O+MNA2O


C     CALCULATE SELECTION QUOTIENTS Q1,Q2,Q3 .  BY MEANS
C     OF LOGICAL STATEMENTS ASSIGN THE SAMPLE TO THE
C     APPROPRIATE TERNARY SYSTEM (1-5).
C     IF Q2 IS GREATER THAN 1.0 AND THERE IS LESS
C     THAN 10% AL2O3 IN THE ANALYSIS, Q2 IS SET TO
C     1.0.


      Q1 = MAL2O3/MSIO2
      Q2 = MAL2O3/MC
      IF (Q2 .GT. 1. .AND. MAL2O3 .LT. .098)   Q2=1
      Q3 = MC/MSIO2
      SYST = 0
      IF (Q1 .LE. .5 .AND. Q2 .GT. 1. .AND. Q3 .LE. .5)     GO TO 50
      IF (Q1 .LE. .5 .AND. Q2 .LE. 1. .AND. Q3 .LE. .5)     GO TO 60
      IF (Q1 .LE. .5 .AND. Q2 .LE. 1. .AND. Q3 .GT. .5 .AND. Q3 .LE. 1.)
     *                                                      GO TO 70
      IF (Q1 .LE. 1. .AND. Q2 .LE. .5 .AND. Q3 .GT. 1. .AND. Q3 .LE. 2.)
     *                                                      GO TO 80
      IF (Q1 .GT. .5 .AND. Q1 .LT. 1. .AND. Q2 .GT. .5 .AND. Q2 .LE. 1.
     *    .AND. Q3 .GT. .5 .AND. Q3 .LT. 2.)                GO TO 90


C     IF THE SAMPLE IS NOT IN TERNARY SYSTEMS 1-5 ASSIGN A VALUE OF 0
C     - GO TO THE NEXT SAMPLE.


      SYST = 0
                                GO TO 100


C     SUBROUTINE UP(1-5) CALCULATES THE THREE MAJOR
C     COMPONENTS OF THE TERNARY SYSTEM.


50    CALL UP1 (A,MC,MAL2O3,MSIO2,AN,AL2O3,FEO)
      SYST = 1


C     IF THERE IS LESS THAN 2% FREE AL2O3 IN SYSTEM 1
C     CALCULATE VALUES FOR BOTH SYSTEMS 1 AND 2.


      IF (SYST .EQ. 1 .AND. AL2O3 .LT. 2.)  SYST=6
      IF (SYST .EQ. 6)                      GO TO 60
                                            GO TO 100
60    CALL UP2 (A,MC,MAL2O3,MSIO2,AN,SIO2,FEO,FAY)
      IF (SYST .NE. 6) SYST=2
                                            GO TO 100
```

```
      70  CALL UP3 (A, MSIO2, MCAO, AN, WO, FEO)
          SYST = 3
          GO TO 100
      80  CALL UP4 (A, MSIO2, MCAO, WO, GEH, FEO)
          SYST = 4
          GO TO 100
      90  CALL UP5 (A, AN, GEH, FEO)
          SYST = 5
     100  IF (SYST .EQ. 6)  WRITE (6, 1030)
          IF (SYST .NE. 6)  WRITE (6, 1020) SYST


          WRITE (6, 2000) (NAME(J), J=1, 10)
          WRITE (6, 2010)
          WRITE (6, 2020)
          WRITE (6, 2030)
          WRITE (6, 2040)
          IF (FE3O4 .NE. 0.0)  WRITE (6, 2050) FE3O4
          WRITE (6, 2060) ((A(I, J), I=1, 2), J=1, 10), C(1), C(2), (A(I, 11), I=1, 2),
         *            C(3), C(4), (A(I, 12), I=1, 2), C(5), C(6), (A(I, 13), I=1, 2),
         *            C(7), C(8), (A(I, 14), I=1, 2), C(9), C(10), (A(I, 15), I=1, 2),
         *            C(11), C(12), (A(I, 16), I=1, 2)
          WRITE (6, 2040)
          WRITE (6, 2070) SUMMA, SUMMB
          IF (SYST .EQ. 0)    WRITE (6, 3000) Q1, Q2, Q3
          IF (SYST .NE. 6)   WRITE (6, 2080) SYST
          IF (SYST .EQ. 6)  WRITE (6, 2090)


          SUBROUTINE ROUND CONVERTS THE VALUES OF THE THREE
          TERNARY SYSTEM COMPONENTS TO INTEGERS WITH A SUM OF
          100%.


          IF (SYST .EQ. 2 .OR. SYST .EQ. 6)  ELA=AN(2)
          IF (SYST .EQ. 4)    ELA = GEH

          IF (SYST .EQ. 1 .OR. SYST .EQ. 3 .OR. SYST .EQ. 5)  ELA=AN(1)

          IF (SYST .EQ. 1)    ELB = AL2O3
          IF (SYST .EQ. 6 .OR. SYST .EQ. 2)  ELB=SIO2
          IF (SYST .EQ. 3 .OR. SYST .EQ. 4)    ELB=WO
          IF (SYST .EQ. 5)    ELB = GEH

          IF (SYST .EQ. 6 .OR. SYST .EQ. 2)  ELC=FEO(2)
          IF (SYST .NE. 6 .AND. SYST .NE. 2)  ELC=FEO(1)

      CALL ROUND(ELA, ELB, ELC)


          IELA = IFIX(ELA)
          IELB = IFIX(ELB)
          IELC = IFIX(ELC)

          IF (SYST .EQ. 1)  WRITE (6, 3010)  Q1, AN(1), IELA, Q2, AL2O3, IELB, Q3,
         *                           FEO(1), IELC
          IF (SYST .EQ. 6)  WRITE (6, 3020)  Q1, AN, IELA, Q2, AL2O3, SIO2, IELB,
         *                           Q3, FEO, IELC, FAY
          IF (SYST .EQ. 2)  WRITE (6, 3030)  Q1, AN(2), IELA, Q2, SIO2, IELB, Q3,
         *                           FEO(2), IELC, FAY
          IF (SYST .EQ. 3)  WRITE (6, 3040)  Q1, AN(1), IELA, Q2, WO, IELB, Q3,
         *                           FEO(1), IELC
          IF (SYST .EQ. 4)  WRITE (6, 3050)  Q1, GEH, IELA, Q2, WO, IELB, Q3,
         *                           FEO(1), IELC
          IF (SYST .EQ. 5)  WRITE (6, 3060)  Q1, AN(1), IELA, Q2, GEH, IELB, Q3,
         *                           FEO(1), IELC


          IF (SYST .EQ. 6)  WRITE (6, 3070)
          IF ( Q2 .EQ. 1.)  WRITE (6, 3080)


      CALCULATE BASICITY

          BASN =(MFEO+MMNO+MMGO+MCAO+MBAO+MK2O+MNA2O)/(MSIO2+MAL2O3)
          WRITE (6, 4000) BASN


      CALCULATE VISCOSITY

          VISC = (A(2, 1)+A(2, 3)+A(2, 4)+A(2, 7)+A(2, 8)+A(2, 9)+A(2, 10))/
         *       (A(2, 5)+A(2, 6))
          WRITE (6, 5000) VISC
          WRITE (6, 110)

     120  READ (5, 1000) NAME
          IF (NAME(1) .EQ. '0000')    GO TO 999
```

```fortran
      IF (NAME(1) .NE. '0000')      GO TO 6
  200 WRITE (6,6000)
  999 CALL EXIT


 1000 FORMAT (     10A4)

          FOR NEW DATA USE FORMAT (10A4) *****

 1010 FORMAT (12A4)
 1015 FORMAT (16F5.2)
 1020 FORMAT (40X, 'TERN.SYST. ',I1/)
 1030 FORMAT (40X, 'TERN.SYST. 2(1)',/)
 2000 FORMAT (5X,10A4//)
 2010 FORMAT (17X, 'WT. %',12X, 'WT. %')
 2020 FORMAT (15X, 'ORIGINAL',9X, 'REDUCED')
 2030 FORMAT (15X, 'ANALYSIS',9X, 'ANALYSIS')
 2040 FORMAT (15X,8('-'),9X,8('-'))
 2060 FORMAT (5X, 'FEO   ',7X,F5.2,12X,F5.2//5X, 'FE2O3',7X,F5 2,12X,F5.2//
     *        5X, 'MNO  ',7X,F5.2,12X,F5.2//5X, 'MGO  ',7X,F5 2,12X,F5.2//
     *        5X, 'AL2O3',7X,F5.2,12X,F5.2//5X, 'SIO2 ',7X,F5.2,12X,F5.2//
     *        5X, 'CAO  ',7X,F5.2,12X,F5.2//5X, 'BAO  ',7X,F5.2,12X,F5.2//
     *        5X, 'K2O  ',7X,F5.2,12X,F5.2//5X, 'NA2O ',7X,F5.2,12X,F5.2//
     *        5(5X,2A4,4X,F5.2,12X,F5.2//),5X,2A4,4X,F5.2,12X,F5.2)
 2050 FORMAT (50X, 'FE3O4 = ',F5.2)
 2070 FORMAT (5X, 'TOTAL',6X,F6.2,11X,F6.2//)
 2080  FORMAT (5X, 'SELECTION QUOTIENTS',8X, 'SYSTEM ',I1/5X,19('-'),
     *         8X,8('-'))
 2090 FORMAT (5X, 'SELECTION QUOTIENTS',8X, 'SYSTEM 1',8X, 'SYSTEM 2',/
     *         5X,19('-'),8X,8('-'),8X,8('-'))
 3000 FORMAT (22X, 'ERROR IN DATA OR MEASUREMENT-NO VALUES CALCULATED',/
     *         5X, 'Q1 = ',F9.5/5X, 'Q2 = ',F9.5/5X, 'Q3 = ',F9.5/)
 3010 FORMAT (5X, 'Q1 = ',F9.5,11X, 'AN   = ',F7.3,'(',I2,
     *         ')'/5X, 'Q2 = ',F9.5,11X, 'AL2O3= ',F7.3,'(',I2,')'/5X,
     *         'Q3 = ',F9.5,11X, 'FEO  = ',F7.3,'(',I2,')'//)
 3020 FORMAT (5X, 'Q1 = ',F9.5,11X,
     *         'AN   = ',F7.3,2X, 'AN   = ',F7.3,'(',I2,')'/5X, 'Q2 = ',
     *         F9.5,11X, 'AL2O3= ',F7.3,2X, 'SIO2 = ',F7.3,'(',I2,')'/5X,
     *         'Q3 = ',F9.5,11X, 'FEO  = ',F7.3,2X, 'FEO  = ',F7.3,'(',I2,
     *         ')'/46X, 'FAY  = ',F7.3/)
 3030 FORMAT (5X, 'Q1 = ',F9.5,11X, 'AN   = ',F7.3,'(',I2,
     *         ')'/5X, 'Q2 = ',F9.5,11X, 'SIO2 = ',F7.3,'(',I2,')'/5X,
     *         'Q3 = ',F9.5,11X, 'FEO  = ',F7.3,'(',I2,')'/
     *         30X, 'FAY  = ',F7.3//)
 3040 FORMAT (5X, 'Q1 = ',F9.5,11X, 'AN   = ',F7.3,'(',I2,
     *         ')'/5X, 'Q2 = ',F9.5,11X, 'WO   = ',F7.3,'(',I2,')'/5X,
     *         'Q3 = ',F9.5,11X, 'FEO  = ',F7.3,'(',I2,')'//)
 3050 FORMAT (5X, 'Q1 = ',F9.5,11X, 'GEH  = ',F7.3,'(',I2,
     *         ')'/5X, 'Q2 = ',F9.5,11X, 'WO   = ',F7.3,'(',I2,')'/5X,
     *         'Q3 = ',F9.5,11X, 'FEO  = ',F7.3,'(',I2,')'//)
 3060 FORMAT (5X, 'Q1 = ',F9.5,11X, 'AN   = ',F7.3,'(',I2,
     *         ')'/5X, 'Q2 = ',F9.5,11X, 'GEH  = ',F7.3,'(',I2,')'/5X,
     *         'Q3 = ',F9.5,11X, 'FEO  = ',F7.3,'(',I2,')'//)
 3070 FORMAT (5X, '** LESS THAN 2% FREE AL2O3 IN SYST. 1'//)
 3080 FORMAT (5X, '** LESS THAN 10% AL2O3 IN ANALYSIS'//)
 4000 FORMAT (5X, 'BASICITY NO. = ',F6.3/)
 5000 FORMAT (5X, 'VISCOSITY COEFFICIENT = ',F6.3//)
 6000 FORMAT (1H0//15('*'), 'SAMPLE NAME ABSENT',15('*'))
  110 FORMAT (1H1)
      END



      SUBROUTINE UP1  (A,MC,MAL2O3,MSIO2,AN,AL2O3,FEO)
      DIMENSION  AN(2),FEO(2),A(2,16)
      REAL       MC,MAL2O3,MSIO2


      TERNARY SYSTEM 1 : ANORTHITE
                        AL2O3
                        FEO



      XAL2O3 = MAL2O3-MC
      TAL2O3 = (A(2,5)*MC)/MAL2O3
      AN(1)  = A(2,6)+A(2,7)+A(2,8)+A(2,9)+A(2,10)+TAL2O3
      AL2O3  = (A(2,5)*XAL2O3)/MAL2O3
      FEO(1) = A(2,1)+A(2,3)+A(2,4)
      RETURN
      END



      SUBROUTINE  UP2 (A,MC,MAL2O3,MSIO2,AN,SIO2,FEO,FAY)
      DIMENSION   AN(2),FEO(2),A(2,16)
      REAL        MC,MAL2O3,MSIO2,MAC,MSI,MMSI,MFE,MFAY,FAY


      TERNARY SYSTEM 2 : ANORTHITE
                        SIO2
                        FEO
```

```
MAC      = MAL2O3+MC
XSIO2    = MSIO2-MAC
TSIO2    = (A(2,6)*MAC)/MSIO2
AN(2)    = A(2,5)+A(2,7)+A(2,8)+A(2,9)+A(2,10)+TSIO2
RSIO2    = (A(2,6)*XSIO2)/MSIO2
SIO2     = RSIO2
FEO(2)   = A(2,1)+A(2,3)+A(2,4)

CALCULATE FAYALITE CONTENT

MSI=SIO2/60.06
MFE=FEO(2)/71.85
MMSI=MSI*2
IF (MMSI .LT. MFE .OR. MMSI .EQ. MFE)    MFAY=MSI
IF (MMSI .GT. MFE)              MFAY=MFE/2
FAY=MFAY*203.76
RETURN
END


SUBROUTINE   UP3 (A,MSIO2,MCAO,AN,WO,FEO)
DIMENSION    AN(2),FEO(2),A(2,16)
REAL         MDSIO2,MSIO2,MCAO


   TERNARY SYSTEM 3:  WOLLASTONITE
                      ANORTHITE
                      FEO



MDSIO2 = MSIO2/3.
XCAO     = MCAO-MDSIO2
TCAO     = (A(2,7)*MDSIO2)/MCAO
TSIO2    = (A(2,6)*MDSIO2)/MSIO2
WO       = TCAO+TSIO2
TXCAO    = (A(2,7)*XCAO)/MCAO
AN(1)    = A(2,5)+A(2,8)+A(2,9)+A(2,10)+TXCAO+(TSIO2*2)
FEO(1)   = A(2,1)+A(2,3)+A(2,4)
RETURN
END


SUBROUTINE   UP4 (A,MSIO2,MCAO,WO,GEH,FEO)
DIMENSION    AN(2),FEO(2),A(2,16)
REAL         MHSIO2,MCAO,MSIO2


   TERNARY SYSTEM 4 :  WOLLASTONITE
                       GEHLENITE
                       FEO



MHSIO2 = MSIO2/2.
XCAO     = MCAO-MHSIO2
TCAO     = (A(2,7)*MHSIO2)/MCAO
TSIO2    = (A(2,6)*MHSIO2)/MSIO2
WO       = TCAO+TSIO2
TXCAO    = (A(2,7)*XCAO)/MCAO
GEH      = A(2,5)+A(2,8)+A(2,9)+A(2,10)+TXCAO+TSIO2
FEO(1)   = A(2,1)+A(2,3)+A(2,4)
RETURN
END


SUBROUTINE   UP5 (A,AN,GEH,FEO)
DIMENSION    AN(2),FEO(2),A(2,16)


   TERNARY SYSTEM 5:  ANORTHITE
                      GEHLENITE
                      FEO


TC       = (A(2,7)+A(2,8)+A(2,9)+A(2,10))/3
TAL2O3 = A(2,5)/2
TSIO2    = A(2,6)/3
AN(1)    = TC+TAL2O3+(2*TSIO2)
GEH      = (2*TC)+TAL2O3+TSIO2
FEO(1)   = A(2,1)+A(2,3)+A(2,4)
RETURN
END
```

```
      SUBROUTINE  ROUND  (ELA,ELB,ELC,SUM)

5 DELA = ELA - AINT(ELA)
  DELB = ELB - AINT(ELB)
  DELC = ELC - AINT(ELC)
  BIG  = AMAX1(DELA,DELB,DELC)

  IF (BIG .EQ. DELA)  ELA = AINT(ELA) +1
  IF (BIG .EQ. DELB)  ELB = AINT(ELB) +1.
  IF (BIG .EQ. DELC)  ELC = AINT(ELC) +1.

  IF (DELA .EQ. DELA .EQ. DELC)  ELA = AINT(ELA) +1

  SUM = AINT(ELA) +AINT(ELB) +AINT(ELC)

  IF (SUM .LT. 100. )  GO TO 5
  RETURN
  END
```

TIMNA30 26. CRYST

	WT. % ORIGINAL ANALYSIS	WT. % REDUCED ANALYSIS
FEO	7. 50	7. 87
FE2O3	0. 00	0. 00
MNO	38. 00	39. 87
MGO	0. 00	0. 00
AL2O3	2. 70	2. 83
SIO2	40. 20	42. 18
CAO	6. 90	7. 24
BAO	0. 00	0. 00
K2O	0. 00	0. 00
NA2O	0. 00	0. 00
CU	0. 80	0. 00
P2O5	3. 00	0. 00
S	0. 50	0. 00
TIO2	0. 00	0. 00
ZN	0. 00	0. 00
PB	0. 00	0. 00
TOTAL	99. 60	100. 00

SELECTION QUOTIENTS   SYSTEM 2
--------------------   --------
Q1 =   0. 03957   AN   = 19. 497(19)
Q2 =   0. 21527   SIO2 = 32. 759(33)
Q3 =   0. 18382   FEO = 47. 744(48)
                             FAY = 67. 699

BASICITY NO. = 1. 097

VISCOSITY COEFFICIENT = 1. 221

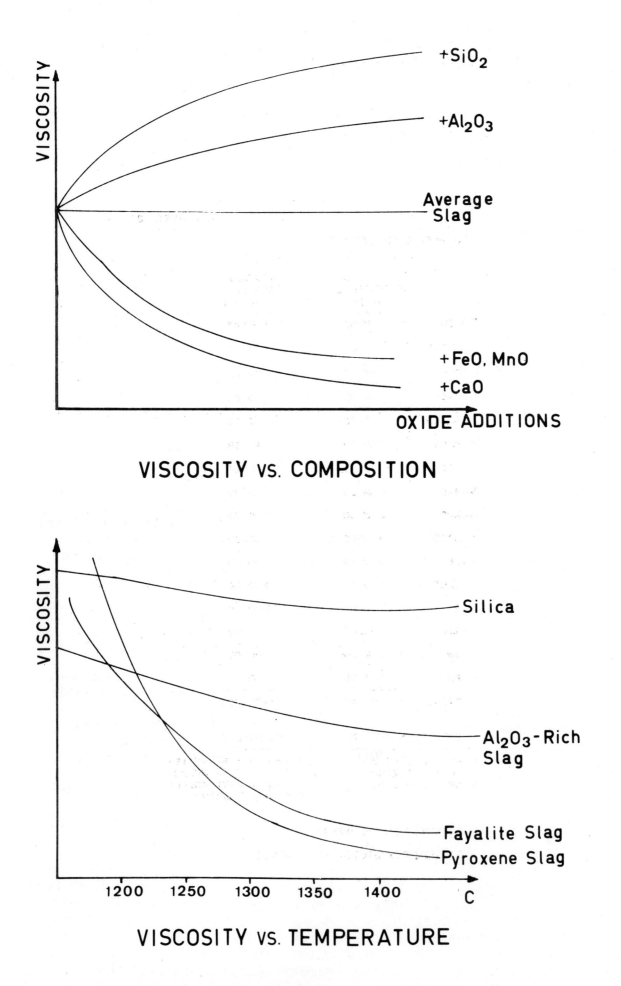

Figure 6: Relation between slag viscosity vs. composition and vs. temperature (schematically).

APPENDIX 2

RELATION BETWEEN VISCOSITY, TEMPERATURE AND SLAG COMPOSITION

The separation of metal from slag is essentially controlled by the viscosity of the slag. The viscosity in turn is influenced by temperature and by the chemical composition of the slag. These relations are schematically shown in figure 6. As the experimental determination of the viscosity of silicate systems at higher temperatures (ie. e. in the region 1100 to 1400°C) is difficult and tedious, we have attempted to derive a mathematical expression to correlate the viscosity of silicate systems, approaching in composition ancient metallurgical slags, with temperature and chemical composition.

According to Hellbrügge & Endell (1941), the viscosity of iron-blast furnace slags can be related to a so-called viscosity coefficient, i. e. the quotient K obtained by dividing the sum of the viscosity-reducing oxide components of a slag, by the sum of the viscosity-increasing components. This equation is identical to the one used for calculating the basicity number (cf. appendix 1) with the only difference that the oxide components are entered in weight percentages. Calculation of K requires the sum of $RO + R_2O + SiO_2 + Al_2O_3$, as given by the experimental analytical data, to be converted to one hundred weight percent. The computer program described in appendix 1 also carries out this calculation. The value of K for slags in general varies approximately between 0. 5 and 2. 5; increasing K signifies reducing viscosity.

Taking the experimentally determined viscosities of modern copper slags from Mansfeld, Germany, as published by Endell and coworkers (1932/33) and those of modern lead slags by Endell, Thielsch & Wens (1934), we calculated the appropriate K-values after "reduction" of the analytical data as mentioned above. As the viscosities of the two groups of slags were given for temperatures between 1100 and 1400°C, we were thus able to submit the dependant variable: viscosity and the independant variables: K and temperature T to a step-wise multiple regression analysis. The following mathematical relation was obtained for η, i. e. the viscosity in Poise ($gcm^{-1}sec^{-1}$):

$$\log \eta = 26.46 - 23.67 \frac{T}{1000} - 3.12 K + 3.84 \left(\frac{T}{1000} \right)^3 - 0.59 K^2$$

The agreement between observed values for η and those calculated with the above formula is reasonably good to justify use of this equation to calculate the viscosity of slags of similar composition for K-values ranging from 0. 5 to 2. 5 within the temperature range 1100 to 1450°C. Figure 7, derived from the above equation for various K-values permits the rapid estimation of viscosities for iron silicate slags, provided their K-values are known from their chemical composition. Within the temperature range of 1150 to 1250°C the viscosity of slags with K-values between 0. 75 and 1. 25 is reduced by a factor of 50! Therefore, any improvement in process technology that increases the temperature of the slag bath within the furnace only moderately has a significant influence on the viscosity of the slag and consequently the velocity and degree of metal/slag separation.

As the K-values rely entirely on the chemical composition of the slags, a word of warning is necessary: The equation used to calculate the K-values is only valid for RO- and R_2O- constituents of silicates or alumosilicates. In slags with separate oxide minerals, e. g. fayalites mixed with spinels - as is often the case in ancient slags - the K-values calculated from the chemical analysis only, disregarding the mineral content of the slag under consideration, are obviously too high, thereby giving a viscosity η which is unrealistically low. This is just another argument to characterise ancient slags by both chemical composition and mineral content.

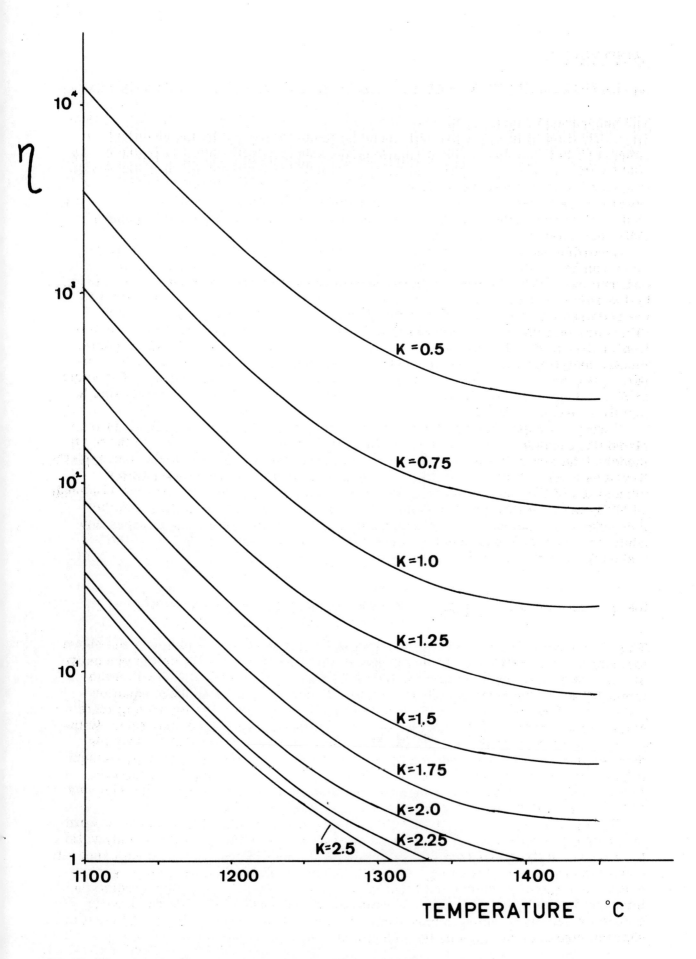

Figure 7: Viscosity vs. temperature in relation to viscosity coefficients.

Acknowledgements

Beno Rothenberg, Tel Aviv, has kindly permitted the use of unpublished chemical analyses of copper slags from Sinai and Timna, carried out for him some years ago by Alexandru Lupu and his coworkers at the Technion, Haifa. The phase analyses were made at the Degussa Research Laboratories, Wolfgang near Hanau, W. Germany. Thanks are due to the Board and Management of this company for the use of their laboratory facilities. Sandra Zacharias, Vancouver, has been of great help in rewriting the computer program (cf. appendix 1) and in preparing the figures. Many stimulating discussions with Peter Wincierz, Metallgesell-schaft A.G., Frankfurt/Main, W. Germany, have helped to find answers to a variety of problems. The financial support of the Stiftung Volkswagenwerk has enabled us to continue and extend our work on the aspects of early copper smelting.

REFERENCES

Bachmann, H.G. 1978. Schlacken: Indikatoren archäometallurgischer Prozesse. In Mineralische Rohstoffe als kulturhistorische Informationsquelle (H.W. Hennicke Ed) Hagen/Germany: Verlag des Vereins Deutscher Emailfachleute, pp. 66-103.

Bachmann, H.G. 1980. On the Identification of Slags from Archaeological Sites. Occasional Publication, Institute of Archaeology, University of London (forthcoming).

Carriveau, G.W. 1974. Application of Thermoluminescence Dating Techniques to Prehistoric Metallurgy. In Application of Science to the Dating of Works of Art (W.J. Young Ed) Boston/Mass.: Museum of Fine Arts, pp. 59-67.

Carriveau, G.W. 1978. Dating of "Phoenician" Slag from Iberia Using Thermoluminescence Techniques. Masca Newsletter, 10, 2 pages (no numbering).

Endell, K., Mühlensiepen, W. and Wagenmann, K. 1932. Metall und Erz 29, pp. 368-375.

Endell, K., Thielsch, W. and Wens, C. 1934. Metall und Erz 31, pp. 353-357.

Hellbrügge, H. and Endell, K. 1940-41. Archiv Eisenhuttenwesen 14, pp. 307-315.

Lupu, A. 1970. Metallurgical Aspects of Chalcolithic Copper Working at Timna (Israel). Bulletin Historical Metallurgy Group 4, pp. 21-23.

Lupu, A. and Rothenberg, B.   1970.   The Extractive Metallurgy of the Early Iron Age Copper Industry in the 'Arabah, Israel. Archaeologica Austriaca 47, pp. 91-130.

Rosenqvist, T.   1974.   Principles of Extractive Metallurgy. Tokyo: McGraw-Hill, Kogakusha Ltd.

Rothenberg, B.   1972.   Timna, Valley of the Biblical Copper Mines. London: Thames and Hudson.

Rothenberg, B.   1978.   Excavations at Timna Site 39, a Chalcolithic Copper Smelting Site and Furnace and its Metallurgy. Archaeo-Metallurgy, IAMS Monograph No. 1, pp. 1-15.

Tite, M. S.   1972.   Methods of Physical Examination in Archaeology. London: Seminar Press.

Wagner, G. A.   1979.   personal communication.

INVESTIGATIONS ON COPPER ORE, PREHISTORIC COPPER SLAG AND COPPER INGOTS FROM SARDINIA

U. ZWICKER[1], P. VIRDIS[2] AND M. L. CERUTI[3]

1. Universität Erlangen-Nürnberg, Lehrstuhl Werkstoffwissenschaft (Metalle), BRD
2. Università degli Studi di Cagliari, Istituto di Chimica Applicata e di Metallurgia, Cagliari-Sardinia/ITALY
3. Università di Cagliari, Facoltà di Lettere, Cagliari-Sardinia/ITALY

Abstract

Copper ore from Funtana Raminosa, slag samples from prehistoric copper smelting places at Nurallao and samples from seven copper ingots from different sites in Sardinia were compared by XRF-, spectrographic, metallographic and microprobe analysis. A slag product of a laboratory reduction of the copper ore with charcoal without any other addition showed the same phases in the metallographic and microprobe analysis as the prehistoric slag from Nurallao and the slag incorporated in two of the ingots. These two ingots had a high oxygen content. In the copper of the matrix of these ingots no cobalt could be detected but in the incorporated slag cobalt was present. Apparently the oxidising condition of the smelting furnace has transferred the cobalt from the ore and from the metal into the slag during a final oxidizing smelting process. The possiblility of a local production of "oxhide" ingots in Sardinia is discussed.

Keywords: COPPER, SLAG, INGOTS, BRONZE AGE, TRADE, PROVENANCE, ANALYSIS, METALLURGY, METALLOGRAPHY.

INTRODUCTION

On the island of Sardinia in at least 12 places pieces or whole ingots called "oxhide" ingots were found (figure 1), (Ceruti 1978). Five of these ingots are already published (Pigorini 1895; 1904; 1905, Vodret 1935; 1959 and Lilliu 1958). On samples from four of these ingots a very precise metallurgical and analytical investigation was performed in another investigation (Balmuth et al. 1976). To find out whether the copper ore and prehistoric slag from a mining and a smelting place in the centre of the island could have been the origin of these blister copper ingots, samples of copper ore from Funtana Raminosa, of slag from Nurallao - a prehistoric smelting place - and samples from seven copper ingots were investigated by X-ray Fluoresence spectrometry (XRF), spectrographic and microprobe analysis. To compare these results with laboratory investigations the ore was reduced by charcoal and the copper and slag produced were investigated in the same way. The ore and the prehistoric slag were heated at 1250°C for half an hour together with copper of high purity and charcoal to find the diffusion of trace elements from the ore or from the slag into the copper.

Figure 1

INVESTIGATION OF THE COPPER ORE FROM FUNTANA RAMINOSA

The different pieces of ore (collected by M. L. Ceruti and P. Virdis 1976) consisted either of pyrite, pyrrothite and chalcopyrite or of iron-oxide. Besides one oxidised pyrite ore all the pieces were ferromagnetic because of their pyrrothite content. One larger piece (ca. 50 grams) of copper ore was investigated in detail. The results of the XRF-analysis are summarized in table 1, those of the spectrographic analysis in table 2 and those of the microprobe analysis in table 3. The elements arsenic, cobalt, lead, manganese, zinc and silver could be detected by spectroscopic analysis. This ore (A) was very inhomogeneous (figures 2 and 3) and consisted mainly of ferromagnetic pink-yellow iron-sulphide (pyrrothite) (P1, P2, P10), areas of golden coloured chalcopyrite (P5, P8) and other dark inclusions. The chalcopyrite contained zinc (P6) or manganese (P3). The pyrrothite may dissolve Co and Ni (P7), iron-oxide (P14) and may contain tungsten (P9, P11, P12). Calcium-oxide can dissolve iron, copper and manganese (P4 and P13).

Pieces of yellow colour (A1, B1, C1) showed no Co in the spectrographic analysis of table 2, whilst the pink coloured pieces (A2, B2, C2) contained Co. The ferrous ore H contained the highest amount of arsenic. Manganese, lead, zinc and silver and traces of tin and gold were also observed in some of these ore pieces.

As was shown in former investigations (Tylecote et al. 1977, Zwicker et al. 1978, Kashima et al. 1978) arsenic from ore or copper matte can be dissolved by copper produced during the smelting process. In 5 grams copper of high purity which was in contact with 10 grams of the ore from Funtana Raminosa and 2.5 grams charcoal at 1250°C for half an hour traces of arsenic were detected by spectrographic analysis (table 2). The micrographs of the copper matte produced by this experiment and of the copper in contact with this matte are shown in figures 4-6. Only sulphides (P1, P3, P4, P6, P7, P9) and oxides (P2, P5, P8, P10, P11) were observed. One oxide inclusion (P8) contained cobalt and one sulphide particle (P9) contained zinc. Sulphide was precipitated in the form of an eutectic between copper and copper sulphide (figure 5).

REDUCTION OF THE ORE BY CHARCOAL

Ten small pieces (ca. 0.3 - 1 grams) of the ore consisting of 9 pieces of mainly chalcopyrite (A1 type) and 1 piece of mainly pyrrothite (A2 type) were reduced by charcoal with air and/or with oxygen. By varying the reaction time the ore was more or less reduced to different reaction products which were investigated by microprobe analysis. Table 4 summarizes the results from these reactions. Short reaction time produced only oxides of copper and iron besides sulphides and silicates. After a reduction time of more than 5 minutes copper was produced, the zinc was dissolved in a Cu-Fe-Zn-oxide and in a Fe-Mn-Cu-Zn-silicate. Figure 7 shows a micrograph of the product from the shortest reduction time (0.2 minutes with compressed oxygen) of ore A1. In P1 and P4 Cu-Fe-oxide, in P2 Fe-Cu-Zn-oxide, in P3 Cu-Fe-Zn-sulphide and in P5 Fe-Cu-sulphide were observed. If the reaction time with compressed oxygen was increased to 1.5 minutes some droplets of metallic copper were produced. By microprobe analysis (figures 8, 9 and 10) metallic copper (P1 and P8), Cu₂O (P5 and P7) and Cu-Fe-oxide (P6) were observed as inclusions within the copper. In the slag Fe-Mn-Ca-Cu-oxide (P2) and Fe-Mn-Cu-oxide (P3) and Fe-Mn-Ca-Cu-K-silicate (P4) could be identified. After a reaction time of 3 minutes with compressed air the microprobe analysis shows (figure 11) that some unreacted silica (P3) was still present besides Fe-Cu-sulphide (P1) and Cu-Fe-Zn- (P2) and Fe-Cu-Zn-sulphide (P4). In the micrographs of figures 12 and 13 after a 5 minutes reduction time with compressed air still many sulphides (Cu-Fe-sulphide, P7 and P8), some metal (P1 and P2) besides a small amount of oxides (Cu-Fe-oxide, P1) can be observed in Fe-Cu-Zn-oxide (P4 and P5) and in the silicious slag (P6). After the same reaction time but with compressed oxygen the amount of metallic copper was increased.

In the agglomerated copper shown in figures 14, 15 and 16 primary copper solid solution is precitpitated within the eutectic reaction between copper solid solution and Cu_2O (P5). The metallic copper in the slag (figure 15) contains a primary crystallization of Cu_2O (P1) in the eutectic reaction of copper solid solution and Cu_2O. In the slag, Cu-Fe-oxide (P4) and Fe-Cu-Mn-oxide (P3) was found within Cu-Ca-Fe-Mn-K-silicate (P2). The long reaction time of 10 minutes with compressed air with ore A1 produced a high content of metallic copper (figures 17 and 18). The blister copper (P1) contains Cu_2O (P2). In the slag Fe-Cu-Zn-oxide (P3), and Cu-Fe-oxide (P5) was precipitated in Cu-Fe-Zn-Mn-K-silicate (P4). The same reaction time with ore type A2 which was followed by a 1 minute reaction with oxygen to get a higher reaction temperature, produced a small amount of metallic iron (figure 19). The microprobe analysis of the main components of the slag shows in P1 a Fe-Mn-Cu-Ca-Al-K-silicate and in P2 metallic copper and iron containing a high amount of Co and Ni.

On the end of the silica tube by which the compressed air was blown into the charcoal, an oxide layer containing zinc-oxide and a small amount of Mn- and As-oxide (table 1) was observed.

INVESTIGATION OF SLAG

Four pieces of slag (collected by M.L. Ceruti and P. Virdis, near Nurallao near Funtana Raminosa in 1956) were analysed by XRF analysis (table 1) and six pieces were investigated by spectrographic analysis (table 2). Because of the high silica content only the trace metals Ag, Mn, Pb and Zn could be detected. All the slag pieces were ferromagnetic. Two of these pieces were further investigated by micrographic examination and microprobe analysis (no. D and F). The micrographs of one of the slag pieces (no. D) which was heavily corroded on the surface are given in figures 20-26. The microprobe analysis shows Cu-Fe-sulphide with a different content of these three elements in P2, P3, P4, P6, P8, P12 and P15, Cu-Fe-Zn-sulphide in P1 and in P5. Cu-Fe-Zn-sulphide containing Pb and/or As was found in P7, Cu-Fe-Zn-silicate in P10, Fe-Cu-silicate containing As and/or Pb in P9 and Ti-Cu-Al-Fe-V-silicate in P14. The second slag piece (no. F), the micrographs of which are given in figures 27-30, shows in P1 and P5 Cu-Fe-sulphide, in P7 Fe-Cu-sulphide, in P2 and P6 Cu-Fe-Zn-sulphide, in P11 Cu-Fe-Zn-Co-sulphide with an uncertain amount of Pb, in P12 a Fe-Cu-Co-Ni-sulphide, in P10 a mixture of silicate and sulphide, in P8 and P9 Fe-Cu-Zn-Pb-silicate, in P4 Fe-Zn-Cu-Mn-Ca-K-silicate and in P3 a similar silicate containing in addition Al.

INVESTIGATION OF THE COPPER INGOTS

In a summary description of "oxhide" ingots "cushion-like" ingots were classified as type I and "oxhide" ingots as type II. All the ingots investigated in this report were classified as type II (Buchholz 1958; 1959; 1966). These ingots are numbered 1-7 upon their finding spots from the south to the north of Sardinia (figure 1). In table 5 all the ingots are described their gas content judged from blistering, the weight and the impressed or inscribed signs and the numbers of other investigations (Buchholz 1959; 1966, Bass 1967, Balmuth et al. 1976). The broad waisted ingots (no. 4 and 6) showed copper oxide and the narrow waisted ingots (no. 1, 2, 3, 5 and 7) showed sulphide inclusions. Therefore the ingots no. 4 and 6 are described first.

BROAD WAISTED INGOTS NO. 4 AND 6

The broad waisted ingot no. 4 (Central Sardinia, Serra Ilixi) (Pigorini et al. 1895; 1904; 1905, Vodret 1935; 1959), no. 3 of Balmuth and Tylecote (Balmuth et al. 1976) is shown in figure 31. It has a double arm impression in the midst of the blister side. In the swarf from the sawing procedure lime was found as an inclusion, probably from the furnace wall, besides a small amount of ferromagnetic slag. The spectrographic analysis showed, in agreement with the former investigations (Vodret 1959, Balmuth 1976), the impurities As, Co, Pb, Sb and Ti. But no Bi, Ni or Zn could be detected in the bulk copper sample. Pieces of slag were incorporated in the copper matrix and by microprobe analysis (figures 32-35 and table 3) it was found that a small amount of Ni, As, Co, Fe and Al was dissolved by a copper matte particle (P1). In another part of this microsection primary crystals of iron oxide (wustite) containing Al, Co and Cu (P6 and P7) were surrounded by Al-Mg-Ca-silicate which contained Co, Cu and Fe (P8). In the copper matrix after a primary crystallization of the copper solid solution a Cu_2O-Cu eutectic has solidified (P5). This ingot was smelted under final oxidizing conditions because the elements As, Ni and Co were dissolved as oxides by the slag. Therefore only traces of Co could be detected by spectrographic analysis (table 2) in the metal and no Co in the solvent extraction by concentrated nitric acid. The highest amount of Co was detected in the ferromagnetic part of the swarf which probably included a large amount of slag.

The other presumed broad waisted ingot no. 6 (Central Sardinia, Teti-Abini) no. 4 of Balmuth and Tylecote (Balmuth et al. 1976) is very fragmentary, indeed only one handle exists (figure 36). The broad waist can be assumed to be similar to ingot no. 4. Ingot no. 6 was found in a bronze hoard. The swarf contained the largest amount of ferromagnetic slag inclusions of all samples investigated. The spectrographic analysis showed such a low content of Co that it could not be detected by the normal method of spectrographic investication (in agreement with the former investigation of Balmuth and Tylecote) (Balmuth et al. 1976). Only in the ferromagnetic part of the swarf was a small amount of Co detected. Zn was only present in the non-ferromagnetic swarf whilst the content of As varied in the different samples (table 2). In the secton for the metallographic investigation and microprobe analysis a large area of slag inclusions was detected (figures 37-40 and table 3). The matrix contained many Cu_2O-inclusions (P2, P13). Some areas were near eutectic. The copper was smelted under final oxidizing conditions. Cu-Fe-oxide (P7), Cu-Fe-Al-oxide (P1, P8 and P9), Fe-Cu-Ti-oxide (P10) and Cu-Fe-Ti-Co-oxide (P3) were the primary crystals of silicate with low copper content (P11) and high copper content (P12) in the slag in which the glassy eutectic matrix of various silicates (P4, P5, P14, P15) contained a higher content of Co. The copper oxide primary crystals contained no cobalt (P17) whilst the iron oxide particles contained cobalt (P16). Only one phase contained sulphide (P6). It can be seen from this ingot and the incorporated slag that the cobalt content of the ore and matte was transferred to the slag by an oxidizing atmosphere during the smelting of the blister copper.

SMALL WAISTED INGOTS NOS. 1; 2; 3; 5 AND 7.

The piece of ingot no. 1 (Capoterras Piscines, Southcoast, not published) has a sign on the top (blister side) which is the same as in ingot no. 2 (figure 41) from Serra Ilixi. The sign could be from the linear A, linear B, Semitic or Phoenician script. In the spectrographic analysis (table 2) Co and As were detected in all samples. In one sample Bi was detected. In the micrograph of the sample corroded areas were observed (figures 42 and 43). The microprobe analysis (table 3) showed a large amount of sulphides. Also sulphide was present (P2) in the interior of the corroded area which consisted mainly of copper oxide with a small amount of Al (P1). In the copper matrix (P3) Cu-Co-Pb-sulphide was precipitated in P4 and Fe-Co-Cu-Ni-Zn-sulphide in P5 and Cu-Ca-silicate in P6.

Ingot no. 2 (Central Sardinia, Serra Ilixi), (Pigorini 1895; 1904; 1905, Vodret 1935; 1959) no. 2 of Balmuth and Tylecote (Balmuth et al. 1976) (figure 41) contained As, Bi, Co, Ti, Mn, Sb, Zn and traces of Pb, Cr and Sn at the limits of detection in agreement with the former investigations (Vodret 1959, Balmuth et al. 1976). In the swarf some ferromagnetic slag particles were present which contained Mn (table 2). The microprobe analysis of the section (figures 44-46) in table 3 showed that sulphides (P4) were present in a large amounts in this ingot also. Some of the copper sulphides contained Co and Pb (P1, P2), others only a very small Fe, Ni, Zn content, but a high Co content (P6, P10). In the entrapped slag the copper-silicate contained chlorine (P7).

The narrow waisted ingot no. 3 (Central Sardinia, Serra Ilixi) (Pigorini 1895; 1904; 1905, Vodret 1935; 1959) no. 1 of Balmuth and Tylecote (Balmuth et al. 1976) had a round hole on the top of the blister side (figure 47) and a sign of a double axe and a star or modified double axe incised into the bottom side of the solidified ingot. The spectrographic analysis showed in agreement with former investigations (Vodret 1959, Balmuth et al. 1976) higher amounts of As and Ag, traces of Co, Pb, Sn, Zn and in addition Cr. In the swarf there were only a few ferromagnetic slag particles which contained a larger amount of Zn (table 2). In the metallographic and microprobe analysis (figures 48-51, table 3) one could see that near the surface there were slag particles and Cu-Fe-oxide without Co (P9) and a small amount of cobalt (P1) and many different sulphides. Also copper chloride was present (P12). The copper matrix of this ingot contained many sulphides of copper (P6, P8, P10). In some of these sulphides a large amount of cobalt and a small amount of Ni and Zn was observed (P3 and P4). In the sulphide of P11 some Co was present. The Fe-sulphide of P7 contained Cu, Co and Zn. One sulphide particle (P14, P15) of the copper matrix (P5) was investigated in detail (figure 51). There was an increase of the Fe- and Co-content and a lower content of sulphur in the outer area (P15) of the particle compared with the central copper sulphide area (P14). In a Cu-Fe-Co-sulphide area a low sulphur content was observed in P2. The silica of P13 contained a small amount of Cu, Al and Fe.

The piece of the ingot no. 5 (Central Sardinia, Isili-Valenze, not published) is similar to the piece of ingot no. 6. No photograph was available. The spectrographic analysis of this sample (table 2) showed As, Bi and Co as major constituents with Pb as a trace and Sn at the limits of detection. The swarf contained only few ferromagnetic slag particles. In the microprobe analysis of the section shown in figures 52 and 53 and table 3 sulphides (P2 and P3), arsenites (P6) and copper oxide (P1) were observed. Co and Fe were found in one of the sulphide particles investigated (P7). Heavy oxidation (P4) had occurred from the surface into the copper matrix, yet the sulphides in this area (P3) were not oxidized.

The small waisted ingot no. 7 (North of Sardinia, Ozieri-S. Antioco di Bisareio), (Vodret 1959), (figure 54) seems to have in an area of the top of the blister side a similarly impressed mark of a double arm as the broad waisted ingot no. 4 (figure 31). In addition a Γ sign on another area was impressed. In the swarf only few ferromagnetic slag particles could be detected. By spectrographic analysis (table 2), besides As, Co and Bi, traces of Sb and Sn were observed. Ni and Zn could not be detected. In the section used for metallographic and microprobe analysis (figure 55, table 3) a Cu-rich oxide particle (P2) and a silicate particle (P6) were observed. This ingot also contained a large amount of sulphides. There were many trace elements included in the sulphides which were precipitated in the copper matrix (P5).

SUMMARY AND DISCUSSION

Many ingots have been found in Sardinia in private collections and by new excavations as is shown in the map of figure 1. Samples from seven ingots were investigated. Five narrow waisted ingots showed sulphide inclusions in the copper matrix. The other two broad waisted ingots contained Cu_2O-Cu eutectic areas but no sulphides in the metallic copper. These sulphide-free ingots were apparently smelted finally in a oxidizing atmosphere for a much longer time than the oxide-free ingots. By the oxidizing reaction the Ni and most of the Co of the copper matrix was transferred into oxide which was dissolved by the slag. Therefore the metallic Ni and Co content of these oxidized ingots was much smaller than in the copper ingots which still contained sulphide particles. The same ore may therefore give a different impurity pattern of Ni and Co if the conditions of the final smelting process are altered.

The composition of the slag found in some of the seven ingots was the same as in the reaction products from the laboratory smelting of the ore from Funtana Raminosa. This ore can be easily smelted to copper. Therefore one can assume that no arduous transportation of copper ingots to the copper rich island of Sardinia may have occurred after Levantine or later Phoenician trading ports had been established in Sardinia. Spectroscopic investigations of the many copper ingots found in the Cape Galedonia bronze age ship wreck (Bass 1967) showed that in the average the Ni-content of these ingots is higher than that of the ingots from Sardinia.

In the Nurage Anastasi at Tertenia besides an ingot fragment a piece of ceramic was found which is believed to be Late Mycenaean coloured ceramic (Ceruti 1978). It shows the connection with the Aegean Sea. This connection of Sardinia with the Aegean Sea is supposed to start in the second half of the third millennium B.C. (Lilliu 1967) and went on up to Phoenician times. Further investigations on ore, slag and copper ingots are necessary to gain an overall picture of Bronze Age production of copper ingots in Sardinia in the Western Mediterranean and in Cyprus and Anatolia in the Eastern Mediterranean. The sometimes similar composition of the ore and of the slag in Sardinia and Cyprus makes a precise decision on the impurity pattern more difficult than it was thought in the beginning of these investigations. The presence of the elements cobalt and nickel can only be expected if there are sulphides present in the copper ingots, which could be found by metallographic and microprobe analysis. This may be a new basis for further investigations.

Acknowledgements

The kind help of Prof. R.F. Tylecote and Dr. P.T. Craddock on reading the manuscript is gratefully acknowledged.

REFERENCES

Balmuth, M.S. and Tylecote, R.F. 1976. Ancient Copper and Bronze in Sardinia: Excavation and Analysis, Journal of Field Archaeology 3, pp. 195-201.

Bass, G.F. 1967. Cape Galedonia Bronze Age Ship Wreck, Transactions of the American Philosophical Society 57, Part 8, pp. 1-174.

Buchholz, H.G. 1958. Der Kupferhandel des 2. vorchristlichen Jahrtausends im Spiegel der Schriftforschung Minoica, Festschrift 80. Geburtstag J. Sundwall: Berlin pp. 92-115.

Buchholz, H.G. 1959. I. Abhandlungen; Keftiubarren und Erzhandel im zweiten vorchristlichen Jahrtausend, Prähistorische Zeitschrift XXXVII, pp. 1-40.

Buchholz, H.G. 1966. TALANTA; Neues über Metallbarren der ostmediterranen Spätbronzezeit, Schweizer Münzblätter 16, pp. 58-72.

Ceruti, M. L.   1978.   Personal communication.

Kashima, M.,   1978.   Transactions of the Japanese Institute of Metals
Eguchi, M. and   19, pp. 152-158.
Yazuwu, A.

Lilliu, G.   1958.   Coittolo inciso prenuragico dalla grotta sarda
                                  di San Michele d'Ozieri, Archaeologia Classica
                                  X Rom, pp. 192, plate LXIV.

Lilliu, G.   1967.   In Sardinien "Frühe Randkulturen des Mittel-
                                  meeres, Korsika-Sardinien-Balearen-Iberer",
                                  Holle-Verlag.Baden-Baden, pp. 29-94.

Pigorini, L.   1895 XXI Pani di rame provenieuti dall'Egeo scoperti a
                         1904 XXX Serra Ilixi in provincia di Cagliari, Bulletino di
                         1905 XXXI Paletnologia Italiana.

Tylecote, R. F.,   1977.   Partitioning of trace elements between the ores,
Ghaznavi, H. A. and   fluxes, slags and metal during the smelting of
Boydell, P. J.   of copper, Journal of Archaeological Science 4
                                  pp. 305-333.

Vodret, F. L.   1935.   Ricerche chimiche sui bronze preistorici della
                                  Sardenga, V. Congress Nat. Chimica Rom,
                                  pp. 108.

Vodret, F. L.   1959.   Sui bronzi preistorici dell'epoca nuragica, 3.
                                  Seminar Facolta Di Science Della Universita Di
                                  Cagliari, Nr. 3/4 XXIX, pp. 1-26.

Zwicker, U. and   1978.   Investigation on the distribution of metallic
Goudarzloo, F.   elements in copper slag, copper matte and copper
                                  and comparison with samples from prehistoric
                                  smelting places, 18th.International Symposium
                                  on archaeometry and archaeometric Prospection,
                                  Bonn, pp. 360-375.

Table 1: X-ray fluorescent spectrometry (XRF) of ore, slag and copper ingots from Sardinia (without background radiation)

Sample	Cu Imp./min.	Fe Imp./min.	Pb/As Imp./min.	Mn Imp./min.	Zn Imp./min.	Ag Imp./min.
Ore Funtana Raminosa piece A not polished	40.080	80.510	1.180	1.250	2.240	1.210?
Ore piece A, other side	21.600	47.000	n.d.	820	1.800	4.750
Ore piece A, metallographic specimen	60.710	51.280	n.d.	1.020	4.620	n.d.
Ore A2 metallographic specimen golden coloured area A1	58.000	61.850	n.d.	n.d.	6.160	n.d.
Ore Funtana Raminosa piece A2	69.690	69.500	n.d.	440	6.380	n.d.
Ore A2 metallographic specimen pink coloured area	45.000	54.640	n.d.	620	6.230	n.d.
Ore Funtana Raminosa piece B	67.220	60.320	n.d.	530	4.820	n.d.
Ore Funtana Raminosa piece C	48.450	66.560	2.210 Pb	600	8.070	n.d.
Ore Funtana Raminosa piece H	n.d.	87.000	1.370 As	550	510?	n.d.
Slag (D) piece	20.250	50.850	1.700 Pb	700	11.120	n.d.
Slag (E) piece	5.180	30.590	900 Pb	n.d.	3.670	n.d.
Slag (E) piece, other side	n.d.	20.400	750 Pb	n.d.	2.630	n.d.
Slag (E) piece	9.320	83.050	2.520 Pb	1.250	8.280	n.d.
Slag (F) piece, other side	28.000	87.460	18.300 Pb	n.d.	19.070	n.d.
Slag (F) piece	n.d.	12.620	n.d.	n.d.	9.50?	n.d.
Slag (G) piece	n.d.	n.d.	n.d.	n.d.	950?	n.d.
Copper Ingot no. 1, piece	273.920	650	n.d.	n.d.	n.d.	n.d.
Copper ingot no.1, metallogr. spec.	248.700	900	n.d.	n.d.	n.d.	n.d.
Copper ingot no. 2, piece	197.810	1.200	1.200	n.d.	n.d.	n.d.
Copper Ingot no. 2, met. specimen	186.400	600	800	n.d.	n.d.	n.d.
Copper Ingot no. 2, piece	383.010	400	1.600	n.d.	n.d.	n.d.
Copper Ingot no. 3, met. specimen	310.470	1.500	1.170	n.d.	n.d.	n.d.
Copper ingot no. 3, piece	362.710	500	1.600	n.d.	n.d.	n.d.
Copper ingot no. 4, piece	345.800	1.500	1.200	n.d.	n.d.	n.d.
Copper ingot no. 4, met. specimen	138.000	350	n.d.	n.d.	n.d.	n.d.
Copper ingot no. 5, met. specimen	206.610	350	n.d.	n.d.	n.d.	n.d.
Copper ingot no. 5, swarf?	263.060	1.650	5.320 Pb	800?	n.d.	n.d.
Copper Ingot no. 6, met. specimen	216.410	34.290	n.d.	n.d.	n.d.	n.d.
Copper Ingot no. 6, met. specimen	96.560	15.570	n.d.	n.d.	n.d.	n.d.
Copper ingot no. 7, piece	n.d.	550	n.d.	n.d.	n.d.	n.d.
Copper Ingot no. 7, met. specimen	n.d.	700	n.d.	n.d.	n.d.	n.d.
Product on silica tube from reactions	n.d.	n.d.	1.470 As	700	960	n.d.

Table 2: Spectrographic analysis (Part 1)

No	Sample	Co	Ni	As	Fe	Si	Zn	Au	Mg	Pb	Sb	Mn	Sn	Ca	Al	Cu	others
79/10	ore A	nd	nd	nd	nd	+	nd	nd	++	nd	nd	nd	nd	++	nd	MC	
103/12	ore A$_1$ (yellow)	nd	nd	tr	MC	ni	+++++	nd	+++++	++	nd	tr?	nd	MC	tr	MC	Ag+++, Ti+
91/13	ore A$_2$ (pink)	+++	nd	nd	MC	tr?	+++	tr	++	++	nd	tr?	tr?	++	tr	MC	Ag +++, Ti +
103/16	ore A$_2$	tr	nd	tr	MC	ni	+++++	nd	+++	+	nd	+++	nd	+++++	nd	MC	Ag+++, Ti+
103/10	ore B$_1$	nd	nd	tr	MC	ni	+++++	nd	+++	++	nd	++	nd	+++++	tr	MC	Ag+++, Ti tr
103/11	ore B2	++	nd	nd	++++	ni	+	nd	+++	tr	nd	+++	nd	+++	tr	MC	Ag ++
103/13	ore C2	+++	nd	nd	+++	ni	tr	nd	++	tr	nd	+++	nd	++	nd	MC	Ag +
103/14	ore C1	nd	nd	tr	MC	ni	+++++	nd	+++	++	nd	+++	nd	MC	tr	MC	Ag+++, Ti+
103/6	ore H	nd	nd	nd	+++	ni	nd	nd	++	nd	nd	tr?	nd	++	nd	nd	
103/7	ore H	nd	nd	+++	MC	ni	nd	nd	++++	tr?	nd	nd	nd	MC	++	nd	Ti tr
103/8	ore H	nd	nd	+	+++++	ni	nd	nd	++++	nd	nd	tr	nd	++	+	nd	Ti tr
103/9	ore H	tr	nd	++	MC	ni	nd	nd	++++	nd	nd	++	nd	+++++	+	nd	Ti tr
106/1	Red. 1250 o C	nd	nd	tr	MC	tr	tr?	nd	+	++	+++?	++	+++	++++	nd	MC	Ag++++, Ti tr?
80/5	Red. 2,5 minO$_2$	nd	nd	nd	MC	++++	nd	nd	++++	nd	ni	++++	nd	+++++	+	MC	Ag tr, Ti tr
80/4	Red. 5 minO$_2$	nd	nd	nd	MC	++++	nd	nd	++++	nd	ni	++++	nd	+++++	++	MC	Ag tr? Ti +
80/3	Red. 10 minO$_2$	+	nd	nd	MC	++	MC	nd	++++	+++	ni	++++	ni	MC	+	MC	Ag MC, Bi tr? Ti tr

Table 2: Spectrographic analysis (Part 2)

NO	Sample	Co	Ni	As	Fe	Si	Zn	Au	Mg	Pb	Sb	Mn	Sn	Ca	Al	Cu	Others
79/2	Slag D	nd	nd	nd	+++	+++	+	nd	+++++	++	nd	+++	nd	+++++	nd	MC	Ag+, Cr tr?, Ti+
103/1	Slag D	nd	nd	nd	+++	ni	tr	nd	+++	++	nd	+	nd	+++++	+	MC	Ti tr, Ag+++
103/2	Slag D	nd	nd	nd	++++	ni	tr	nd	++++	+	nd	nd	nd	++++	++	MC	Ti tr, Ag++++
79/1	Slag E	nd	nd	nd	+++	+++	tr	nd	+++++	tr	nd	++	nd	+++++	++	MC	Ag tr
103/15	Slag E	nd	nd	nd	++++	ni	nd	nd	++++	tr	nd	+	nd	+++	+	+++++	Ti tr, Ag ++
103/4	Slag F	nd	nd	nd	+	ni	nd	nd	tr	tr	nd	nd	nd	+++	nd	MC	
103/5	Slag F	nd	nd	nd	+++++	ni	tr	nd	++++	++	nd	++	nd	++++	+	MC	Ti tr
80/1	Slag G	nd	nd	nd	+	++++	nd	nd	+++	nd	nd	nd	nd	++	tr	tr	
80/2	Slag H	nd	nd	tr?	+	tr	nd	nd	+	nd	nd	nd	nd	tr	tr	tr	
No.	Ingot no.	Co	Ni	As	Fe	Si	Zn	Au	Mg	Pb	Sb	Mn	Sn	Ca	Al	Cu	Others
30	1 bulk	tr	nd	++	++	++	nd	nd	+	tr	tr?	+	nd	++++	++	MC	
27	1 bulk	+	nd	++	+++	+++	nd	nd	+	nd	nd	nd	nd	+++	nd	MC	
28	1 bulk	++	nd	++	++++	+++	tr	nd	++	tr?	tr?	tr?	nd	+++++	++	MC	Bi +
110/19	1 bulk	+	nd	++	+++	++	nd	nd	+	nd	nd	nd	nd	++++	++	MC	Ti +
79/7	1 solvent	nd	nd	tr	tr	ni	nd	nd	+	nd	nd	nd	nd	++	++	MC	Ti tr
110/6	1 swarf	tr	nd	tr	tr	ni	tr	nd	++	nd	nd	nd	nd	+++	nd	MC	Ti tr, Ag tr
103/21	1 swarf ferromagn.	tr	nd	tr	++	ni	tr	nd	++	nd	nd	+	nd	+++	nd	MC	Ti +, Ag tr?
104/1	1 swarf not ferromagn.	tr	nd	+	+	ni	+++	nd	+	nd	nd	nd	nd	++	nd	MC	Ti +, Ag +

Table 2: Spectrographic analysis (Part 3)

No.	Ingot no.	Co	Ni	As	Fe	Si	Zn	Au	Mg	Pb	Sb	Mn	Sn	Ca	Al	Cu	Others
30	2 bulk	++	nd	++	++	++	nd	nd	++	tr	++	nd	nd	+++	++	MC	Ti ++
27	2 bulk	++	nd	++	++	++	nd	nd	++	tr	nd	nd	tr?	+++	++	MC	Cr tr
28	2 bulk	++	nd	++	++	++	+	nd	++	tr?	nd	+	nd	MC	++	MC	Cr tr₂, Bi +
103718	2 bulk	+	nd	tr	++	++	nd	nd	++	nd	nd	nd	nd	++	+	MC	Ti +, Ag +
7976	2 solvent	+	nd	tr	++	tr	nd	nd	nd	nd	nd	nd	nd	++	nd	MC	Ti tr₂, Ag tr
11075	2 swarf	tr	nd	tr	+	ni	nd	nd	tr	nd	nd	nd	nd	++	nd	MC	Ti tr₂, Ag tr
103720	2 swarf ferro-magnetic	nd	nd	tr	+	ni	tr	nd	tr	nd	nd	nd	nd	nd	nd	MC	Ti tr
103727	2 swarf not ferromagn.	+	nd	tr	++	ni	++	nd	+	+	nd	nd	tr	++	nd	MC	Ti +, Ag tr
30	3 bulk	tr	nd	+	++	+++	nd	nd	++	tr	nd	nd	nd	++	++	MC	Cr tr₂, Ti +
27	3 bulk	nd	nd	tr	++	++	tr	nd	+	nd	nd	nd	tr?	+++	++	MC	
28	3 bulk	nd	nd	tr	++	+++	nd	nd	++	nd	nd	nd	nd	+++	++	MC	Ti tr₂, Ag +
110715	3 bulk	nd	nd	nd	nd	nd	nd	nd	nd	nd	nd	nd	nd	++	tr	MC	Ti tr₂, Ag +
7973	3 solvent	+	nd	tr	++	tr	nd	nd	++	nd	nd	nd	nd	+++	nd	MC	Ti tr₂, Ag +
11072	3 swarf	nd	nd	tr	nd	ni	nd	nd	tr	nd	nd	nd	nd	++	nd	MC	Ti tr
103724	3 swarf ferro-magnetic	tr	nd	tr?	+++	nd	+	nd	+	nd	nd	++	nd	++	nd	MC	Bi +, Ag +++
103735	3 swarf not ferromagn.	+	nd	+	+	nd	++	nd	+	nd	nd	nd	tr?	++	nd	MC	Ti tr, Ag tr
30	4 bulk	tr	nd	++	++	++	nd	nd	++	++	tr	nd	nd	+++	++	MC	Cr tr₂, Ti ++
27	4 bulk	nd	nd	++	++	++	nd	nd	+	nd	nd	nd	nd	+++	tr	MC	
28	4 bulk	tr	nd	++	++	+++	nd	nd	++	tr?	nd?	tr?	nd	MC	+	MC	
110717	4 bulk	nd	nd	tr	nd	tr	nd	nd	tr	nd	nd	nd	nd	++	nd	MC	Ti tr₂, Ag tr?
7975	4 solvent	nd	nd	tr	+	++	nd	nd	++	nd	nd	nd	nd	+++	tr	MC	Ti tr₂, Ag tr
11074	4 swarf	nd	nd	tr	nd	ni	nd	nd	tr	nd	nd	nd	nd	+	tr	MC	Ti tr
103719	4 swarf ferro-magnetic	+	nd	+++	+++	ni	nd	nd	+++	nd	nd	nd	nd	+++	tr	++++	Ti tr
103726	4 swarf not ferronagnetic	nd	nd	+	tr	ni	++	nd	++	+	nd	nd	tr	+++	nd	MC	Ti +, Ag tr

Table 2: Spectrographic analysis (Part 4)

No.	Ingot no.	Co	Ni	As	Fe	Si	Zn	Au	Mg	Pb	Sb	Mn	Sn	Ca	Al	Cu	Others
30	110/8 5 bulk	++		++	++	++	++	nd	+++	nd	++	nd	nd	++	++	MC	Bi +
	110/21 5 bulk	tr	nd	++	+++	ni	nd	nd	nd	nd	nd	nd	nd	+++	+++	MC	Tl tr
	79/9 5 solvent	tr	nd	tr	+++	ni	nd	tr?	++	nd	nd	nd	nd	++	tr	MC	Tl tr, Ag tr
	103/23 5 magnetic ferro	tr	nd	++	nd	ni	nd	nd	tr	nd	nd	nd	nd	++	nd	MC	Tl tr
10473	5 swarf not ferromagn.	+	nd	+	++	ni	nd	nd	+	nd	nd	nd	nd	++	tr	MC	Tl +, Ag tr?
	103/23 5 magnetic	++	nd	++	++	ni	+	nd	++	++	+	+	nd	++++	nd	MC	Tl tr
103/25	6 swarf not ferromagn.	nd	+	+	++	ni	+++	nd	++	+	nd	nd	+	++	nd	MC	Cr tr
103/18	6 swarf ferro	tr	nd	nd	+++	ni	nd	nd	+++	nd	nd	tr	nd	+++	+	+++++	Tl tr
79/4	6 solvent	nd	nd	nd	tr	++	nd	tr?	++++	nd	nd	nd	nd	tr	tr	MC	Cr +, Tl ++
110/16	6 bulk	nd	nd	nd	+++	++	nd	nd	++++	nd	nd	nd	nd	++	nd	MC	
110/3	6 bulk	++	nd	++	+++	+++	nd	nd	+++	nd	nd	nd	nd	+++	+	MC	
28	6 bulk	+	nd	++	+++	++	nd	nd	+++	nd	nd	nd	nd	++++	+	MC	
27	6 bulk	+	nd	+	+++	+++	nd	nd	++	nd	nd	nd	nd	+++	nd	MC	
30	6 bulk	nd	nd	nd	++	+	nd	nd	+++	tr	nd	nd	tr	++	nd	MC	Cr ++, Tl ++, Bi tr
30	7 bulk	++	nd	++	+++	+	nd	nd	+++	tr	+	nd	tr	++	+	MC	Cr tr
27	7 bulk	+	nd	+	++	tr	nd	nd	++	nd	nd	nd	nd	++	+	MC	
28	7 bulk	+	nd	+	++	tr	nd	nd	+	nd	nd	nd	nd	++	+	MC	
110/20	7 bulk	tr	nd	+	++	nd	nd	nd	++	nd	nd	nd	nd	++	nd	MC	Tl tr
79/8	7 solvent	tr	nd	tr	tr	tr	nd	nd	tr	nd	nd	nd	nd	++	nd	MC	Ag tr, Tl tr
110/7	7 swarf	nd	nd	+	nd	ni	nd	nd	+	nd	nd	nd	nd	+++	nd	MC	Ag tr? Tl tr
103/22	7 swarf ferro	+++	nd	+	++	ni	+++	nd	++	nd	nd	nd	tr	+++	nd	MC	Ag ++, Tl
	103/22 7 magnetic	+++	nd	+	++	ni	+++	nd	++	+	nd	nd	tr	+++	nd	MC	Ag +, Tl ++
10472	7 swarf not ferromagn.	++	nd	+	++	ni	+++	nd	+	nd	nd	nd	nd	++	nd	MC	Ag + Tl +

MC = main constituent
++++ → + decreasing content
tr = trace

nd = non detected
ni = not investigated because fixed with silica

Table 3: Microprobe analysis (Part 1)

Fig. no.	Area	Mg	Si	S	Ca	Mn	Fe	Co	Ni	Cu	Zn	Others
	P1	nd	nd	51.0	nd	nd	66.5	nd	nd	nd	nd	nd
	P2	nd	nd	51.4	nd	nd	73.0	nd	nd	nd	nd	nd
	P3	nd	nd	27.4	nd	0.17	51.3	nd	nd	45.6	nd	nd
	P4	1.22	nd	6.1	14.6	2.57	14.6	nd	nd	0.67	nd	nd
2 and 3	P5	nd	nd	3.1	nd	nd	47.0	nd	nd	43.0	nd	nd
	P6	nd	nd	26.2	nd	nd	27.0	nd	nd	43.0	3.97	nd
(Ore	P7	nd	nd	40.6	nd	nd	45.4	0.29	0.15	0.16	nd	nd
Funtana	P8	nd	nd	20.1	nd	nd	35.1	nd	nd	30.3	nd	nd
Raminosa)	P9	nd	nd	nd	nd	nd	70.8	nd	nd	2.11	nd	W 0.21
	P10	nd	nd	13.8	nd	nd	82.2	nd	nd	2.32	nd	nd
	P11	nd	nd	nd	nd	nd	79.7	nd	nd	0.88	nd	Ti 0.61 W 0.19
	P12	nd	nd	nd	nd	nd	70.7	nd	nd	nd	nd	W 0.37
	P13	0.68	nd	nd	16.0	3.62	7.27	nd	nd	nd	nd	nd
	P14	nd	nd	nd	nd	nd	84.7	nd	nd	nd	nd	nd
	P1	nd	nd	13.5	nd	nd	0.34	nd	nd	108	nd	nd
	P2	nd	nd	nd	nd	nd	0.34	nd	nd	111	nd	nd
	P3	nd	nd	5.4	nd	nd	1.49	nd	nd	32.1	nd	nd
4 - 6	P4	nd	nd	7.1	nd	0.26	51	nd	nd	26.9	nd	nd
(Reaction	P5	nd	nd	nd	nd	nd	0.55	nd	nd	93.0	nd	nd
Ore +	P6	nd	nd	14.1	nd	nd	1.82	nd	nd	71.0	0.40	nd
copper	P7	nd	nd	9.3	3.5	nd	4.2	nd	nd	7.1	nd	nd
1250 °C)	P8	nd	nd	nd	nd	nd	90.0	0.72	nd	3.35	nd	nd
	P9	nd	nd	4.8	nd	nd	59.0	nd	nd	20.2	0.69	nd
	P10	nd	nd	nd	nd	nd	1.59	nd	nd	85.0	nd	nd
	P11	nd	nd	nd	nd	nd	1.24	nd	nd	50.3	nd	Pb 25.8 Bi 7.8
7	P1	nd	nd	nd	nd	nd	7.59	nd	nd	82.1	nd	nd
(Reduction	P2	nd	nd	nd	nd	nd	84.7	nd	nd	7.37	0.41	nd
oxygen	P3	nd	nd	15.1	nd	nd	29.3	nd	nd	50.9	0.85	nd
0,2min	P4	nd	nd	nd	nd	nd	15.5	nd	nd	79.5	nd	nd
Ore A1)	P5	nd	nd	19.8	nd	nd	58.7	nd	nd	8.38	nd	nd

Table 3: Microprobe analysis (Part 2)

Fig. no.	Area	Mg	Si	S	Ca	Mn	Fe	Co	Ni	Cu	Zn	Others
8 - 10 (reduction 1,5 min oxygen ore A1)	P1	nd	nd	nd	nd	nd	2.23	nd	nd	107.5	nd	nd
	P2	nd	nd	nd	0.46	0.63	7.2	nd	nd	0.36	nd	nd
	P3	nd	nd	nd	nd	0.55	84.1	nd	nd	0.16	nd	nd
	P4	nd	10.0	nd	15.0	0.75	19.6	nd	nd	0.26	nd	K 0,18
	P5	nd	nd	nd	nd	nd	nd	nd	nd	98.6	nd	nd
	P6	nd	nd	nd	nd	nd	24.8	nd	nd	52.2	nd	nd
	P7	nd	nd	nd	nd	nd	nd	nd	nd	82.1	nd	nd
	P8	nd	nd	nd	nd	nd	0.34	nd	nd	107.5	nd	nd
11 (reduction 3 min air ore A1)	P1	nd	nd	18.1	nd	nd	85.0	nd	nd	22.7	nd	nd
	P2	nd	nd	19.1	nd	nd	23.4	nd	nd	82.5	1.27	nd
	P3	nd	55.0	nd	nd	nd	nd	nd	nd	nd	nd	nd
	P4	nd	nd	24.4	nd	nd	61.0	nd	nd	29.2	2.78	nd
12 - 13 (reduction 3 min air ore A2)	P1	nd	nd	5.95	nd	nd	20.2	nd	nd	93.0	nd	nd
	P2	nd	nd	nd	nd	nd	4.76	nd	nd	108	nd	nd
	P3	nd	nd	30.7	nd	nd	62.5	nd	nd	2.79	0.36	nd
	P4	nd	nd	nd	nd	nd	75.0	nd	nd	0.50	0.88	nd
	P5	nd	8.1	nd	nd	nd	59.0	nd	nd	13.8	0.78	nd
	P6	nd	nd	15.4	nd	0.37	54.2	nd	nd	0.78	nd	nd
	P7	nd	nd	nd	nd	nd	6.55	nd	nd	68.5	nd	nd
	P8	nd	nd	14.3	nd	nd	13.3	nd	nd	57.5	nd	nd
14 - 16 (reduction 5 min oxyg. ore A1)	P1	nd	nd	nd	nd	nd	2.06	nd	nd	77.1	nd	nd
	P2	nd	11.0	nd	9.60	0.62	9.00	nd	nd	6.95	nd	K 0.50
	P3	nd	nd	nd	nd	0.42	70.8	nd	nd	2.36	nd	nd
	P4	nd	nd	nd	nd	nd	15.5	nd	nd	70.6	nd	nd
	P5	nd	nd	nd	nd	nd	nd	nd	nd	92.5	nd	nd
17 - 18 (reduct. 10 min air ore A1)	P1	nd	nd	nd	nd	nd	nd	nd	nd	145	nd	nd
	P2	nd	nd	nd	nd	nd	99.0	nd	nd	122	nd	nd
	P3	nd	nd	nd	nd	nd	nd	nd	nd	6.3	1.30	nd
	P4	nd	10.9	nd	nd	0.28	24.2	nd	nd	25.5	1.10	K 0.40
	P5	nd	nd	nd	nd	nd	nd	nd	nd	108	nd	nd
19 (reduct. 10 min air + 1 min O_2 ore A2)	P1	nd	6,8	nd	11.0	0.42	44	nd	nd	0.29	nd	nd
	P2	nd	nd	nd	nd	nd	39	8.9	3.7	57.1	nd	Al 1.29; K 0.25

Table 3: Microprobe analysis (Part 3)

Fig. no.	Area	Mg	Si	S	Ca	Mn	Fe	Co	Ni	Cu	Zn	Others
	P1	nd	nd	1.94	nd	nd	1.14	nd	nd	70.7	0.33	nd
	P2	nd	nd	2.36	nd	nd	0.87	nd	nd	64.1	nd	nd
	P3	nd	nd	2.62	nd	nd	3.22	nd	nd	55.2	nd	nd
	P4	nd	nd	10.7	nd	nd	1.65	nd	nd	66.1	nd	nd
	P5	nd	nd	14.4	nd	nd	0.72	nd	nd	48.0	0.21	nd
20 - 26	P6	nd	nd	16.0	nd	nd	19.6	nd	nd	39.2	nd	nd
	P7	nd	nd	23.8	nd	nd	35.1	nd	nd	17.5	1.64	As or Pb 0.38
	P8	nd	nd	17.0	nd	nd	15.1	nd	nd	40.2	nd	nd
(Slag D)	P9	nd	1.85		nd	nd	32.1	nd	nd	19.0	nd	As or Pb 0.48
	P10	nd	1.25	nd	nd	nd	27.0	nd	nd	13.4	0.34	nd
	P11	nd	4.89	nd	nd	nd	24.8	nd	nd	17.6	0.40	nd
	P12	nd	nd	17.8	nd	nd	24.7	nd	nd	49.1	nd	nd
	P13	nd	nd	nd	nd	nd	0.89	nd	nd	84.6	nd	As or Pb 0.25
	P14	nd	1.08	nd	nd	nd	1.03	nd	nd	20.9	nd	Al 1.08, Ti 31.7 V 1.85
	P15	nd	nd	10.6	nd	nd	1.37	nd	nd	77.1	nd	nd
	P1	nd	nd	16.9	nd	nd	25.5	nd	nd	52.1	nd	nd
	P2	nd	nd	7.40	nd	nd	30.7	nd	nd	30.7	0.80	nd
	P3	nd	11.4	nd	5.28	0.27	19.6	nd	nd	0.44	1.41	Al 1.39, K 0.18,
	P4	nd	5.44	nd	0.50	0.93	49.3	nd	nd	0.23	2.19	K 0.18
27 - 30	P5	nd	nd	11.3	nd	nd	0.75	nd	nd	77.1	nd	nd
	P6	nd	nd	12.0	nd	nd	1.45	nd	nd	66.1	0.41	nd
	P7	nd	nd	21.0	nd	nd	58.7	nd	nd	1.01	nd	nd
(Slag F)	P8	nd	1.79	nd	nd	nd	42.7	nd	nd	8.06	nd	Pb 0.40
	P9	nd	2.57	nd	nd	nd	44.0	nd	nd	8.62	0.28	As or Pb 0.22
	P10	nd	1.12	2.9	nd	nd	31.2	nd	nd	40.3	0.20	As or Pb 0.48
	P11	nd	nd	8.30	nd	nd	29.4	0.21	nd	40.3	nd	As or Pb 0.54
	P12	nd	nd	19.6	nd	nd	68.2	0.32	0.23	2.00	nd	nd

Table 3: Microprobe analysis (Part 4)

Fig. no.	Area	Mg	Si	S	Ca	Mn	Fe	Co	Ni	Cu	Zn	Others
32 - 35	P1	nd	nd	3,82	nd	nd	3,50	3,85	2,06	50,96	nd	Al 0,262, As 2,28
	P2	nd	nd	nd	nd	nd	nd	nd	nd	66,17	nd	nd
	P3	nd	nd	nd	nd	nd	nd	nd	nd	70,64	nd	nd
	P4	nd	1,56	0,737	nd	nd	nd	nd	nd	1,86	nd	Cr 5,5; Pb 5,3; Ti4,5
	P5	nd	nd	nd	nd	nd	41,50	0,29	nd	79,6	nd	Al 4,00
	P6	nd	nd	nd	nd	nd	0,32	nd	nd	0,32	nd	Al 0,83
	P7	nd	nd	nd	nd	nd	57,3	1,56	nd	0,65	nd	Al 1,05
(Ingot no. 4)	P8	0,27	12,9	nd	1,28	nd	20,3	0,66	nd	0,55	nd	nd
37 - 40	P1	nd	nd	nd	nd	nd	0,645	nd	nd	82,06	nd	Al 0,34
	P2	nd	nd	nd	nd	nd	1,41	nd	nd	66,14	nd	nd
	P3	nd	21,86	nd	nd	nd	66,29	0,64	nd	1,12	nd	Al 0,67; Ti 0,54
	P4	nd	22,3	nd	2,64	nd	20,35	nd	nd	0,262	nd	Al 1,59; K 0,32
	P5	nd	1,01	1,66	3,05	nd	17,66	0,192	nd	1,56	nd	Al 1,96; K 0,127
	P6	nd	nd	nd	nd	nd	6,35	nd	nd	70,66	nd	Al 0,483
	P7	nd	nd	nd	nd	nd	4,80	nd	nd	68,12	nd	nd
	P8	nd	nd	nd	nd	nd	2,375	nd	nd	60,69	nd	Al 0,254
	P9	nd	nd	nd	nd	nd	2,64	nd	nd	62,17	nd	Al 0,62
(Ingot no. 6)	P10	nd	13,4	nd	0,301	nd	62,29	nd	nd	2,13	nd	Ti 0,27
	P11	nd	2,06	nd	nd	nd	37,09	nd	nd	1,07	nd	nd
	P12	nd	nd	nd	nd	nd	17,62	nd	nd	32,05	nd	Al 0,172
	P13	nd	13,43	nd	nd	nd	0,625	0,768	nd	82,16	nd	nd
	P14	nd	25,23	nd	0,273	0,137	40,42	nd	nd	0,437	nd	Al 2,00; K 0,605; Ti 0,622
	P15	nd	nd	nd	3,76	nd	18,67	0,17	nd	0,407	nd	Al 0,622
	P16	nd	nd	nd	nd	nd	49,32	nd	nd	0,180	nd	nd
	P17	nd	nd	nd	nd	nd	2,92	2,55	nd	68,17	nd	nd
42 + 43	P1	nd	nd	nd	nd	nd	nd	nd	nd	62,18	nd	Al 1,17
	P2	nd	nd	5,73	nd	nd	nd	nd	nd	57,17	nd	nd
	P3	nd	nd	nd	nd	nd	nd	nd	nd	84,61	nd	nd
	P4	nd	nd	13,20	nd	nd	nd	1,48	nd	57,14	nd	nd
	P5	nd	nd	21,20	0,156	nd	42,78	19,55	2,37	3,39	2,04	Pb 0,529
(Ingot no. 1)	P6	nd	8,23	nd	nd	nd	nd	nd	nd	39,25	nd	nd

Table 3: Microprobe analysis (Part 5)

Fig. no.	Area	Mg	Si	S	Ca	Mn	Fe	Co	Ni	Cu	Zn	Others
44 - 46 (Ingot no. 2)	P1	nd	nd	15,31	nd	nd	nd	0,216	nd	55,17	nd	Al 0,28; Pb 2,29
	P2	nd	nd	14,72	nd	nd	nd	1,76	nd	58,61	nd	As or Pb 1,03
	P3	nd	nd	nd	nd	nd	nd	nd	nd	64,16	nd	nd
	P4	nd	nd	25,4	nd	nd	nd	nd	nd	60,64	nd	nd
	P5	nd	nd	nd	nd	nd	nd	nd	nd	66,17	nd	nd
	P6	nd	nd	nd	nd	nd	49,29	14,17	0,15	3,18	1,12	nd
	P7	nd	13,22	nd	0,277	nd	1,78	nd	nd	33,07	nd	Al 6,40;Cl 20,20
	P8	nd	nd	nd	nd	nd	nd	nd	nd	79,70	nd	nd
	P9	nd	nd	nd	nd	nd	nd	nd	nd	113,6	nd	nd
	P10	nd	nd	nd	nd	nd	74,7	15,5	nd	4,48	1,26	nd
48 - 51 (Ingot no. 3)	P1	nd	nd	nd	nd	nd	0,46	0,236	nd	77,17	nd	nd
	P2	nd	nd	9,0	nd	nd	0,395	0,34	nd	70,64	nd	nd
	P3	nd	nd	4,58?	nd	nd	43,98	18,58	0,29	3,74	1,23	nd
	P4	nd	nd	15,34	nd	nd	43,98	17,58	0,145	2,375	1,16	nd
	P5	nd	nd	nd	nd	nd	nd	nd	nd	77,13	nd	As or Pb 0,28
	P6	nd	nd	16,35	nd	nd	nd	nd	nd	64,16	nd	nd
	P7	nd	nd	6,09	nd	nd	49,32	6,82	nd	2,66	0,21	nd
	P8	nd	nd	26,15	nd	nd	nd	nd	nd	55,19	nd	nd
	P9	nd	nd	nd	nd	nd	0,286	nd	nd	70,69	nd	nd
	P10	nd	nd	21,65	nd	nd	nd	0,417	nd	57,18	nd	Cl 23,91
	P11	nd	nd	37,64	nd	nd	nd	nd	nd	45,31	nd	Al 0,642
	P12	nd	nd	nd	nd	nd	nd	nd	nd	49,21	nd	nd
	P13	nd	64,9	nd	nd	nd	0,13	nd	nd	7,51	nd	nd
	P14	nd	nd	10,6	nd	nd	nd	nd	nd	87,1	nd	nd
	P15	nd	nd	9,50	nd	nd	0,12	0,2?	nd	87,1	nd	nd
52 - 53 (Ingot no. 5)	P1	nd	nd	nd	nd	nd	nd	nd	nd	64,2	nd	nd
	P2	nd	nd	19,5	nd	nd	nd	nd	nd	53,9	nd	nd
	P3	nd	nd	21,05	nd	nd	nd	nd	nd	55,2	nd	nd
	P4	nd	nd	nd	nd	nd	nd	nd	nd	62,2	nd	nd
	P5	nd	nd	0,382	nd	nd	nd	nd	nd	52,2	nd	nd
	P6	nd	nd	nd	nd	nd	nd	nd	nd	49,2	nd	As 3,26
	P7	nd	nd	6,18	nd	nd	42,8	6,98	nd	2,76	nd	nd
55 (Ingot no. 7)	P1	nd	nd	4,30	nd	nd	0,437	0,23	nd	62,17	nd	nd
	P2	nd	nd	nd	nd	nd	0,513	0,177	nd	62,17	nd	nd
	P3	nd	nd	nd	ni	ni	42,77	10,75	0,382	28,50	0,66	nd
	P4	nd	nd	16,37	nd	nd	nd	nd	nd	68,10	nd	nd
	P5	nd	nd	nd	nd	nd	nd	nd	nd	87,1	nd	nd
	P6	nd	2,90	8,15	nd	nd	19,57	3,99	nd	39,2	nd	Al 0,54

Table 4: Reduction of Funtana Raminosa ore by charcoal at temperatures higher than 1200°C (A1 type of ore = low content of Co, A2 type of ore = higher content of Co)

reaction no.	ore type	oxidizing atmosphere	reaction time min	reaction products
R 99	A1	oxygen	0,2	sulfide, Cu-Fe-oxide, Cu-Fe-Zn-oxide, Cu-Fe-Zn-sulfide, Fe-x-silicates (Fig. 7), no Cu
R 173	A1	air	1,5	Ferromagnetic chalco-pyrite with precipitation of Cu, eutectic Fe-silicate (low and high Cu-content
R 174	A1	oxygen	2,0	Ferromagnetic Cu_2O, eutectic Fe-silicate (low and high Cu-content) third silicate phase in primary crystallisation
R 97	A1	oxygen	2,5	Cu pure and with low Fe-content, Fe-Ca-K-Mn-Cu-silicate, Cu-Fe-x-oxide, Cu_2O (Fig. 8 - 10)
R 192	A1	air	3,0 fast cooling	similar to R 191 and R 193
R 193	A1	air	slowly cooled 3,0	similar to R 191 and R 192, not reacted SiO_2, Cu-Fe-Zn-sulfide, Cu-Fe-sulfide (Fig. 11)
R 191	A1	air	5	Cu-precipitation from sulfide, Cu-Fe-oxide, Fe-Cu-Zn-oxide, Fe-Zn-Mn-silicate, Cu-Fe-sulfide (Fig. 12 and 13)
R 98	A1	oxygen	5	Cu, Cu_2O (eutectic), Fe-Mn-K-Ca-Cu-silicate, Cu-Fe-Mn-oxide, Cu-Fe-oxide (Fig. 14 - 16)
R 190	A1	air	10	Cu, Cu_2O, Cu with low Fe-content only in slag, Cu-Fe-Zn-oxide, Fe-Mn-Cu-Zn-silicate (Fig. 17 and 18)
R 283	A2	air oxygen	10 1	Fe + Cu containing Co and Ni, Fe-Mn-Cu-Ca-Al-K-silicate (Fig. 19)

Table 5: Description of the copper ingots

no. of ingot (Fig.)	Type of Buch-holz 13)	no. of Buch-holz 13)	no. of Bass 14)	no. of Tyle-cote 8)	weight kg	gas content (blis-tering)	oxide content micros-copic matrix	sulfide content micros-copic matrix	magne-tic slag in saw dust	impression blister (top) side	incised sign bottom side	max. length cm	min. length cm	max. breadth cm	min. breadth cm	Finding place
no. 1 (41)	II	-	-	-	1,6	++	no	+++	+++	[symbol] (fracture)	?	?	?	?	?	Capoterras Piscines (South Coast)
no. 2 (41)	II	59	11 + 13	2	33,6	++	no	+++	+++	[symbol]	[symbol]	72	47	35	17	Serra Ilixi 1)2)3)
no. 3 (47)	II	60	14 + 15	1	34,0	++	no	+++	+ (+CaO)	[symbol]	[symbol]	64,5	45	34	18	Serra Ilixi 1)2)3)
no. 4 (31)	II(I)	61	12	3	26,4	++++	++	no	+	[symbol]	no	52,5	47,5	33	27,5	Serra Ilixi 1)2)3)
no. 5	II?	-	-	-	3,5	++	no	+++	+	(handle)	?	?	?	?	?	Isili-Valenza
no. 6 (36)	II(I)	-	-	4	0,8	++++	+++	no	+++	(handle)	?	?	?	?	?	Teti-Abini
no. 7 (54)	II (III)	-	16?	-	22,5	++	no	+++	+	[symbol]	no	66	43	39	24	Ozieri-S. Antioco di Bisareo

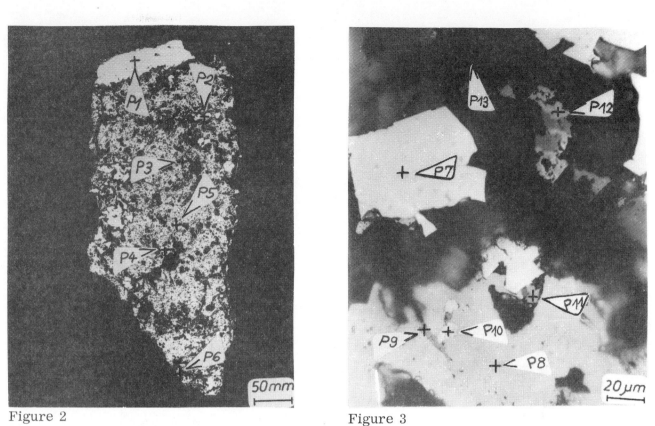

Figure 2

Figure 3

Figures 2 and 3: Micrographs of ore from Funtana Raminosa, Sardinia, and of areas of microprobe analysis

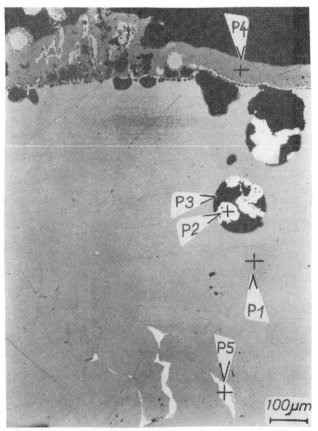

Figure 4:   Microstructure of copper matte

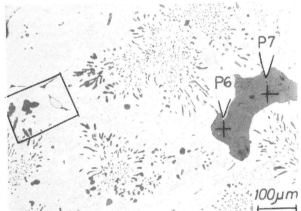

Figure 5: Microstructure of copper

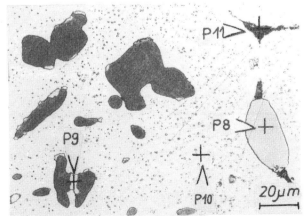

Figure 6: Microstructure of copper

Figures 4 - 6: Reaction of ore with Cu at 1250°C, areas of microprobe analysis

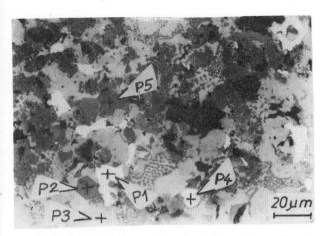

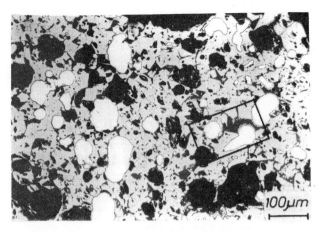

Figure 7: Microstructure of ore A1 reduction 0.2 min oxygen

Figure 8: Microstructure of ore A1 reduction 1.5 min oxygen

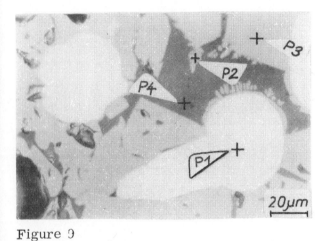

Figure 9

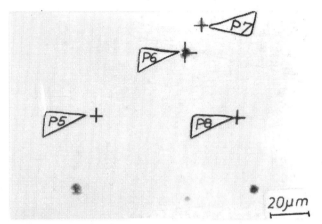

Figure 10

Figures 9 and 10: Microstructure and areas of microprobe analysis of ore A1 reduction 1.5 min oxygen

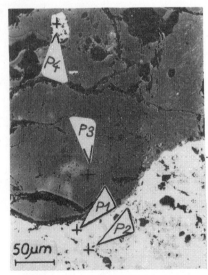

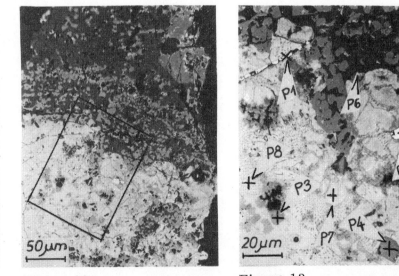

Figure 11: Micrograph and areas of micro- probe analysis reduction 3 min air of ore A1

Figure 12

Figure 13

Figures 12 and 13: Micrographs and areas of microprobe analysis, reduction of ore A1 5 min air

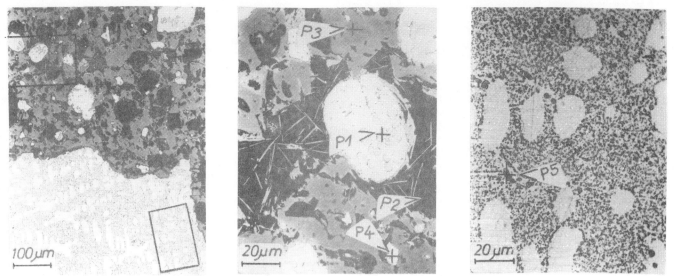

Figure 14   Figure 15   Figure 16

Figures 14 - 16: Micrographs and areas of microprobe analysis reduction 5 min oxygen of ore A1

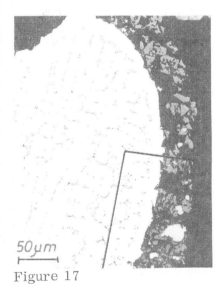

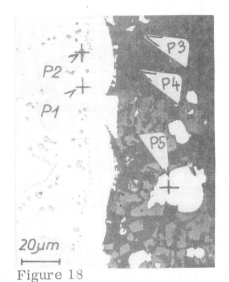

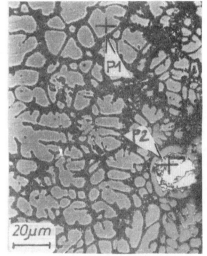

Figure 17   Figure 18   Figure 19 : Micrograph and

Figures 17 - 18: Micrographs and areas of microprobe analysis reduction of ore A1 10 min air

Figure 19 : Micrograph and area of micro- probe analysis reduction of ore A2 10 min air + 1 min oxygen

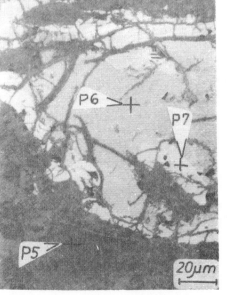

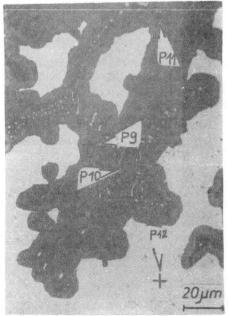

Figure 20   Figure 21

Figures 20 and 21: Micrographs and areas of micro- probe analysis of slag D (oxidized area)

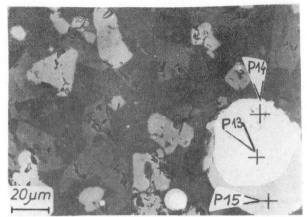

Figure 22

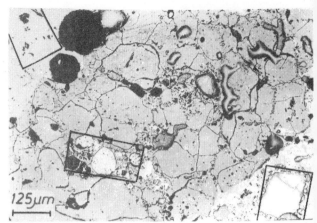

Figure 23

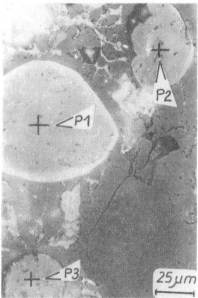

Figure 24

Figure 25

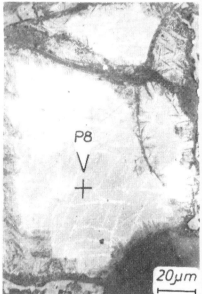

Figure 26

Figures 22 – 26 : Micrographs and areas of microprobe analysis of slag D (not oxidized area)

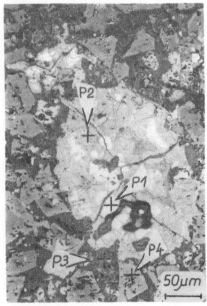

Figure 27

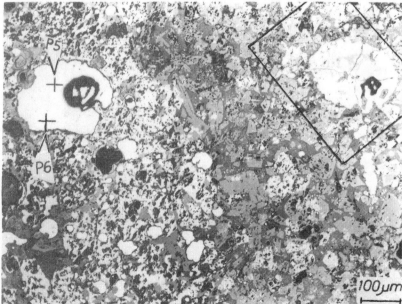

Figure 28

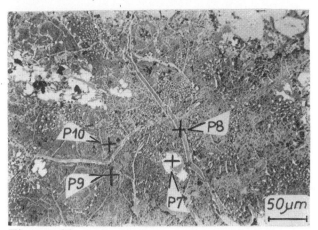

Figure 29

Figure 30

Figures 29 and 30: Micrographs and areas of microprobe analysis of slag F.

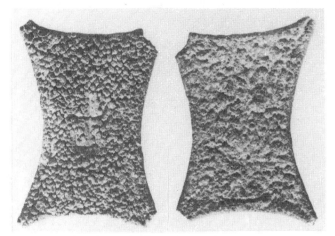

Figure 31: Copper ingot no. 4

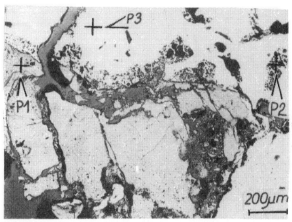

Figure 32

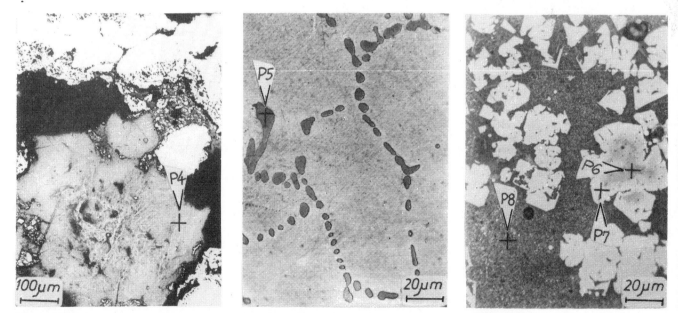

Figure 33

Figure 34

Figure 35

Figures 32 – 35: Micrographs and areas of microprobe analysis of copper ingot no. 4

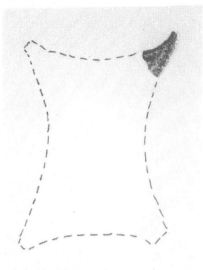

Figure 36:   Copper ingot no. 6

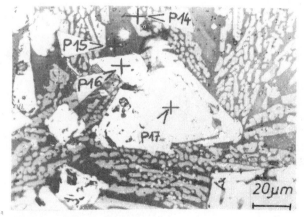

Figure 37

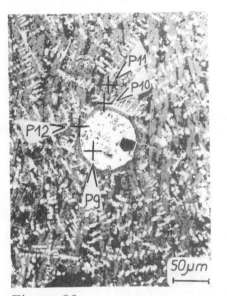

Figure 38

Figures 37 – 40: Micrographs and areas of microprobe
analysis of copper ingot no. 6

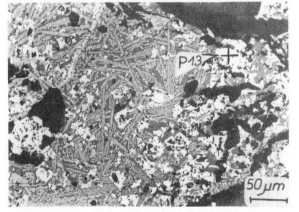

Figure 39

Figure 40

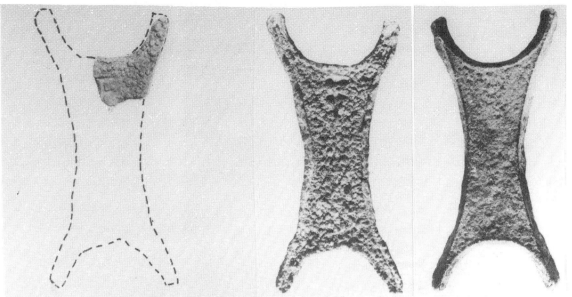

Figure 41: Copper ingot no. 1 (left) and copper ingot no. 2 (right)

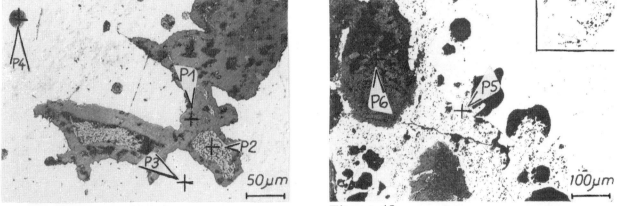

Figure 42

Figure 43

Figures 42 and 43: Micrographs and areas of microprobe analysis of copper ingot no. 1

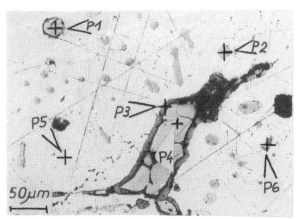

Figure 44

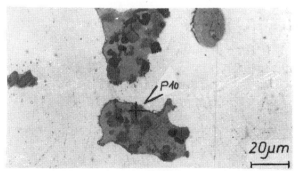

Figure 45

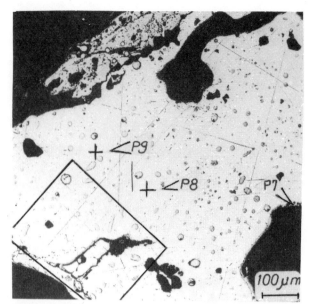

Figure 46

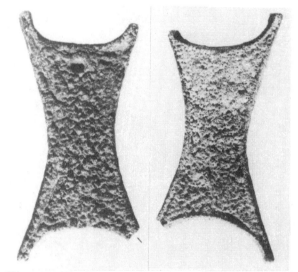

Figure 47 : Copper ingot no. 3

Figures 44 – 46 : Micrographs and areas
of microprobe analysis
of copper ingot no. 2

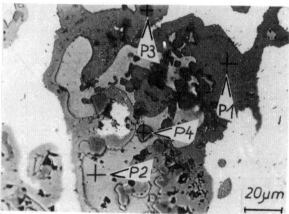

Figure 48

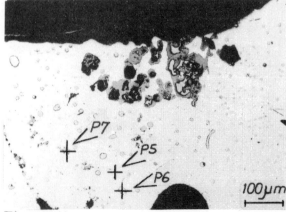

Figure 49

Figures 48 – 51 : Micrographs and areas
of microprobe analysis
of copper ingot no. 3

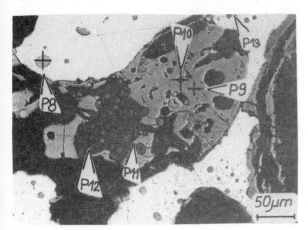

Figure 50

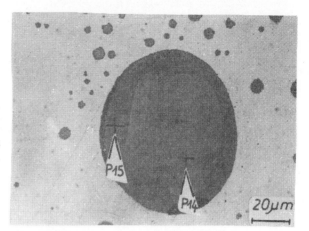

Figure 51

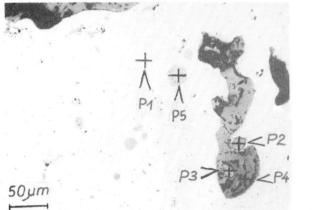

Figure 52

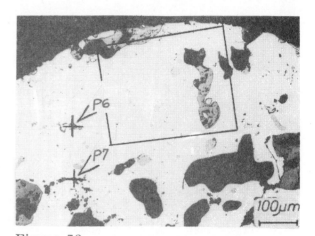

Figure 53

Figures 52 and 53 : Micrographs and areas of microprobe analysis of copper ingot no. 5

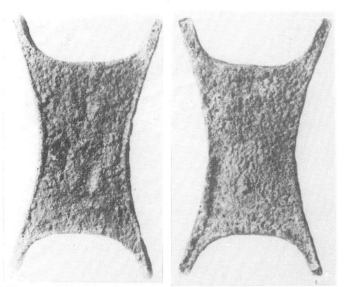

Figure 54 : Copper ingot no. 7

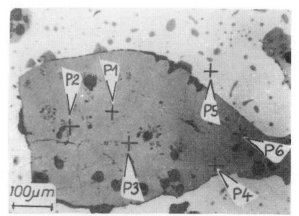

Figure 55 : Micrograph and areas
of microprobe analysis
of copper ingots no. 7

THE COMPOSITION OF COPPER PRODUCED AT THE ANCIENT SMELTING CAMPS IN THE WADI TIMNA, ISRAEL

PAUL T. CRADDOCK

Research Laboratory, British Museum, London WC1B 3DG

Abstract

The analysis of over 300 artifacts of unrefined copper from the Late Bronze Age smelting sites at Timna provide a unique insight into the production and distribution of copper in the Ancient World. Study of the minor and trace metals incorporated in the copper give valuable parameters to the smelting process. The level of these impurities, especially of iron is high enough to affect the working properties of the copper; explaining the concern of ancient authors with types of copper from specific localities.

Although no tin occurs naturally in the area, small amounts of the metal occur in three-quarters of the samples; at present there is no explanation for these small but persistent quantities of tin in the copper.

Keywords: COPPER, MINING, MALACHITE, BRONZE AGE, ANALYSIS, ATOMIC ABSORPTION, SLAG, SMELTING, TRADE, INGOTS, METALLURGY.

INTRODUCTION

The remains of the copper mining and smelting activities in the Wadi Timna are amongst the most extensive and complete yet known from the ancient world and prolonged field work and scientific examination of the remains (Rothenberg 1972, 78), have already probably contributed more to our understanding of the production of copper in antiquity than has the study of any other site. But even here the investigations are far from complete, and the potential of the site is far from exhausted. Indeed, this paper, as the philosopher would have predicted, raises as many new questions as it settles old ones.

The subject presented here is the composition of 318 copper-base artifacts excavated at Timna and presumed to have been made there. Such a volume of metal finds from any archaeological site would be unusual, but from a metal-working site it is unique. The bulk of the material consists of offerings made to a Temple of Hathor (site T. 200) situated a few hundred metres from one of the main smelting camps (T. 30) and dating to the Late Bronze Age, the most intensive period of extraction at Timna. The bulk of the finds are of small roughly made rings, pins and trinkets, but including metal in all stages of production from spills of raw metal in accidental zoomorphic shapes, to finished castings (Plate 1), together with many small offcuts, and fragments. A complete catalogue and description of all the finds, with their composition where applicable, is shortly to be published in a volume devoted to the Temple by B. Rothenberg. The excavations in the mines and smelting camps have also produced copper and bronze tools, although not in such prodigious numbers.

This collection of metal from one area provides a unique opportunity to study the composition of the unrefined metal straight from the smelting furnace and prior to it's removal to artifact production centres where it would be purified and possibly mixed with other metal and scrap, thus masking beyond recognition the relationship of the metal composition to the ore and smelting process. The data from Timna thus allows one to relate the ancient copper composition both to the ore and to the ancient smelting processes as revealed by archaeology and as reproduced in the very valuable smelting experiments using reconstructions of ancient furnace types which have recently been carried out. (Tylecote 1977, Tylecote et al 1977 and 1978).

A second important point on which these Timna analyses provide information is the purity and degree of consistency in composition to be expected for copper from one production centre in antiquity. The highly impure, but characteristic copper that was the final product from Timna shows why ancient authors such as Pliny could regard a metal as having distinct chemical and physical properties depending on where it was produced.

METHOD

The samples were usually in the form of drillings except where the artifact proved too small, when a scraping was taken with a scalpel. The samples were analysed by Atomic Absorption Spectrometry in the normal way (Hughes et al, 1976) but using a graphite furnace attachment for the arsenic, antimony and bismuth determinations. A full description of the sampling procedure will be given together with the complete analytical results for 15 elements in the forthcoming Temple Volume.

DISCUSSION

The ore body exploited in antiquity at Timna is in the 'Middle White Horizon' of the Nubian Sandstone and is a secondary body formed principally of malachite nodules with some cuprite, azurite and chrysocolla (Slatkine 1961). Tylecote et al (1967) reported some chalcocite in the complex ore nodules, and sulphides were also undoubtedly present in the malachite ores, this being confirmed by the presence of sulphides in the ancient metal produced at Timna. This is in contrast to the complete absence of sulphur in the primary chrysocolla deposits which are mined at great depth by the modern Timna works. (Bartura and Wurzburger 1974). In general though the ores smelted in ancient times at Timna were secondary oxides, from which most of the associated metals, such as arsenic, antimony, silver, lead, zinc, nickel and bismuth, had been substantially leached away. Analyses of the ore taken from the ancient workings and smelting camps confirms that the trace metal content is low. (Tylecote et al 1977, Lupu 1970, Field in Rothenberg 1972, and in figure 1 of this work), and this is reflected in the similarly low trace element content in the extracted metal (figure 1).

The copper ores were smelted with acacia charcoal using iron oxide to form a slag with the silica gangue of the copper ore. This iron oxide was also obtained from the Middle White Horizon where it exists as haematite formed from the conversion of fossilised wood.

EXPERIMENTAL SMELTINGS

Tylecote et al (1977) have performed experimental smelts in a small furnace based on an excavated example, F39 ascribed to the Chalcolithic period from Timna. (Rothenberg et al 1978), and it is instructive to compare the composition of the copper from these experiments with that extracted by the smiths of antiquity (see figure 1). Two sets of experiments were carried out, the first in which only one Tuyere was used (1) and the second using two tuyeres (2). With one tuyere, which is typical of Chalcolithic and Bronze Age practise, the temperature in the furnace did not rise high enough, especially in the slag, for a discreet ingot of copper to form, but instead globules and prills of metal remained dispersed in the slag. By

	Cu	Pb	As	Sb	Ni	Ag	Zn	Fe	Mn	Bi
Timna ore used in 1 and 2	23.7	0.045	0.05	nd	0.013	tr.	tr.	7.3	-	nd
Experimental smelt 1		-	0.3	0.02	0.024	0.03	0.025	1.2	(15)	-
Experimental smelt 2		nd	0.2	nd	0.04	-	(50)	2.7	0.07	nd
Timna ore (recent analyses)	21	0.04	0.01	(3)	0.015	(30)	0.12	6.1	0.2	(10)
Ancient Timna copper (average of 318 analyses)		0.36	0.125	0.025	0.035	0.024	0.025	1.5	(15)	0.01

figures in parenthesis are in ppm, all others expressed in %

Figure 1: Analyses of Timna ores, metal and the products of Tylecote's experimental smelts.

using two tuyeres, a much higher temperature was achieved throughout the furnace and a copper ingot formed beneath the slag (see figure 2 p. 307 Tylecote et al 1977). The ore used by Tylecote was Timna malachite, but the flux was a synthetic iron oxide, which was responsible for at least one important difference in composition (see below). Figure 1 shows the composition of the copper ore used by Tylecote and of the copper produced by furnaces (1) and (2), (these are taken from Table 7, p. 313 of Tylecote 1977). Beneath these, for comparison, is an average composition of 6 malachite ores collected from Timna mines and camps by the author, and the average composition for the 318 ancient artifacts analysed. Tylecote (1977) has shown that in the strongly reducing conditions of a furnace for smelting oxide ores, the retention of volatile elements such as zinc, bismuth, antimony and lead is strongly dependent on the temperature. This is clearly shown by comparing smelt (1) with smelt (2) in figure 1, both sets of analyses were performed on the unpurified metal straight from the furnace. The ancient metal would seem to be most similar to that produced in Tylecote's furnace (1) which used only one tuyere, and in which only a somewhat lower temperature was produced, especially in the slag. A result of this is that the copper tended to stay in the slag as discrete globules and prills which had to be mechanically removed by breaking up the cooled slag. The smelting camps at Timna, which are contemporary with the Temple and the metalwork studied here, contain huge heaps of broken slag from which the raw copper globules have been removed. Only with more tuyeres and a higher slag temperature, as in (2), does a separate ingot form beneath the slag. Whilst the molten copper is moving downwards through the slag at the more elevated temperatures it can pick up more iron, as has been demonstrated by Tylecote. The manganese content is clearly associated with the iron rich phase of the metal as would be expected on metallurgical grounds, however the relatively high level of manganese in Tylecote's (2) metal is rather surprising as apparently neither the copper ore or the flux contained detectable manganese (Tylecote et al 1977 Table 1).

The arsenic content is relatively low in the ancient metal compared to that in the experimental smelts (1) & (2). Tylecote et al (1977 p. 314) suggested this could be due to prior roasting or the use of a higher furnace temperature, but this also should have reduced the other volatile trace elements accordingly. A more probable reason is the relatively high arsenic content of the Timna ore used by Tylecote, (500 ppm). The six further analyses recently performed by the author had an average arsenic content of only 100 ppm (see figure 1).

The most striking difference between the ancient and the modern experimental Timna metal is the lead content. The reason for this is not associated with either the ore which is similar or the smelting procedure which by and large gives a very similar product. Instead, it is most probably due to the use of synthetic iron oxide as the flux in the experimental smelting rather than the fossilized tree trunks used in antiquity. Furthermore this explanation is consistent with Rothenberg's (1972) earlier suggestion that the lead content in the metal was coming from the flux not the ore. Analysis of the Timna flux has revealed the presence of lead.

IRON CONTENT

An important aspect convincingly demonstrated both by Tylecote's work and the analyses discussed here is that the inforporation of several per cent of metallic iron into the copper can be regarded as usual. Oxide ores of copper are invariably associated with a predominance of siliceous matter which must be removed during smelting, by the addition of a metal oxide to flux away the silica as a glassy slag. As stated above iron oxides are ideal for this and were widely used in antiquity at Timna and elsewhere; occasionally other metal oxides such as manganese oxide were used, as in the later phase at Timna itself. During the smelting some of the iron oxide flux can itself be reduced to iron metal and become incorporated in the slag. The molten copper within the slag can therefore pick up several percent of this iron quite easily. Now as oxide ores of copper are believed to have been widely exploited in antiquity particularly in the Early Bronze Age (Coghlan 1951), most copper produced at this period should contain concentrations of iron similar to that found in the Timna metalwork. However, inspection of the many thousands of analyses that are now available for bronze from all over the Ancient World shows that, although high iron contents are occasionally reported from nearly all regions and periods, (Cooke and Aschenbrenner 1975) only a minute proportion of the total contain more than traces of iron. Clearly therefore, the iron in the raw copper was usually removed prior to use.

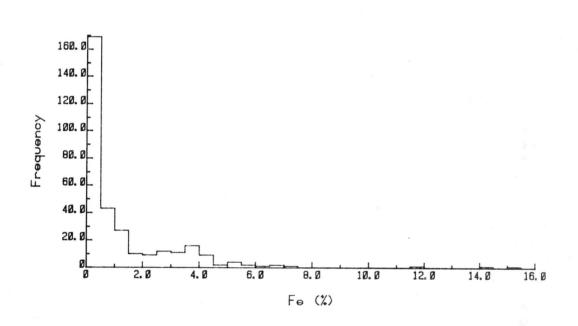

Figure 2 :   Iron content of Timna copper.

Iron does not form a solid solution with copper, and if present in more than traces it is extremely deleterious to the working and use of the copper. Tylecote and Boydell (1978) were able to reduce the iron content of the raw copper quite easily by remelting. The iron then either floats to the top forming an iron rich phase which can be skimmed off or forms a slag with the silica of the clay crucible which once again floats to the surface and can be removed. A unique and totally unexpected feature of the ancient Timna metalwork was the high iron content in the finished bronze alloys as well as the raw metal (figure 2). It is clear that the Timna metalsmiths made no attempt to purify the metal prior to use. This view is reinforced by the relatively high Sb, As, Zn, Ni & Ag contents which would also have been substantially reduced by purification (Tylecote et al 1977 Table 7). In view of Tylecote's work indicating the ease of purification, it is surprising that cast and alloyed metal from Timna are equally rich in iron, since both processes require remelting in a clay crucible . This fact is important in a general context since it shows that iron had to be deliberately removed by the metalsmiths and that the operations of alloying and casting the metal would not alone be sufficient for its adventious removal.

Clearly for the Timna smith the metal obtained by smelting was the final product. This conclusion is supported by the find of a bun ingot in the Wadi Arabah, many kilometres from Timna, which contains about 3 percent of iron, proving that the metal left the smelting camps in this impure state (Rothenberg 1972, page 69). Other hoards of ingots in the Near and Middle East are a good source of comparative material, but few have yet been analysed. Exceptions are the Middle Bronze Age hoard of ingot bars from the Hebron Hills and Har Yeruham, Israel, examined by Maddin and Wheeler (1976). These ingots contained between 0.3 and 1.7% of iron. Similarly of the ten oxhide ingots from off Cape Gelidonya, Southern Turkey, quantitatively analysed 8 contained only traces of iron, one contained between 1 and 2% (Muhly, Wheeler and Maddin 1977), and one contained 10% (Maddin and Wheeler 1974).

The material from smelting camps, and the primary ingots from hoards give us much information about the copper before it reached the artisan's workshops in the cities for purification and alloying. We should expect differences in composition depending on the ore and smelting process. Thus the smelting of carbonate and oxide ores produces an iron rich copper, whereas the copper from a sulphide ore will contain much less iron. Although copper sulphide ores first began to be used in the Bronze Age, oxide and carbonate ores continued in use. For example, the Roman smelting site of Be'Ora adjacent to Timna smelted malachite, and the ingot from there contains 5% of iron. Thus certainly by Roman times a wide variety of ores and smelting procedures were in use, producing raw copper of widely differing compositions. These compositions would often have been characteristic of the process and place of production.

LITERARY SOURCES

Pliny and other ancient authors who discuss copper and other metals often attribute different intrinsic properties to metals derived from different areas, and indicate that some of these are to be preferred to others. Pliny in the opening chapters of Book 34 of his Natural History, (Rackham 1968), devoted to copper, iron and lead, first describes the natural 'types' of copper from various localities and lists them in order of excellence. Clearly Pliny believed, for example, that the copper from Gaul was intrinsically different and better than that from Cordova. To modern minds brought up on the concept of 92 natural elements, and the Law of Constant Composition this seems strange. However, the theoretical treatment of the origin of matter, as stated by Aristotle and based on the four elements fire, air, water and earth, held that metals were a 'vaporous exhalation' formed by the sun's rays upon water. The exhalation was entrapped in the earth, the dryness of which gradually converted it to metal. (Holmyard 1968). Thus there was no intrinsic theoretical reason in the time of Pliny why a metal should have constant properties. Indeed common experience would suggest that the very varied environmental and geological circumstances in which the metal 'grew' should produce a varying product. Pliny says of iron 'There are numerous varieties of iron the first difference depending on the kind of soil or climate'. (Natural History Book 34 chapter 41).

Sometimes, however, Pliny realised that the smelting and purification could be held responsible. Thus in the case of Cyprus copper he states 'Bar copper also is produced in other mines, likewise fused copper. The difference between them is that the latter can only be fused as it breaks under the hammer, whereas bar copper otherwise called ductile copper is malleable, which is the case with all Cyprus copper. But also in the other mines this difference of bar copper from fused copper is produced by treatment; for all copper after impurities have been rather carefully removed by fire and melted out of it becomes bar copper'. (Natural History Book 34, chapter 20). This must be a reference to the presence of sulphides or iron in the metal which would certainly have to be removed before the copper became ductile.

If the metal from some production centres was of as poor a quality as that from Timna, it is small wonder that these coppers were recognised as different from copper from centres such as Cyprus with more rigorous quality control.

TIN CONTENT

Over two thirds of the Timna metalwork contains tin in amounts varying from traces up to 13% (figure 3). Tin bronze had by this time become the usual alloy in use throughout Europe and Western Asia, but the normal range is about 8 to 10% of tin. The tin content in the Timna artifacts is much more variable, and usually much lower than elsewhere. In fact the presence of these very small amounts is baffling.

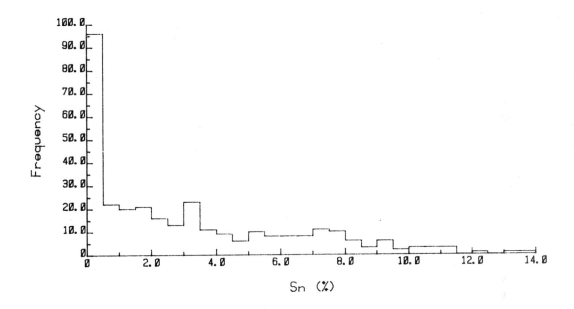

Figure 3: Tin content of Timna copper.

There is a correlation between the various methods of producing wire practised at Timna and the tin content (Oddy in the forthcoming Temple report), although there is no obvious reason why there should be so. However, there is no general correlation between the tin content and the artifact type; spills of metal, pins, tools all contain essentially random amount of tin and it is as if the smiths were unaware of its presence. Yet at Timna this is impossible since the ores do not contain tin, and the geology of the area rules out any local source for the metal. Therefore, the tin in the Timna metalwork must have been added deliberately.

The discovery of two fragments of tin at Timna suggests that it was added in metallic form to the copper. The iron content and general nature of the metalwork further suggests that the copper is primary, and that we are not seeing the very diluted tin content of scrap bronzes after repeated remelting and addition of more copper. If anything there is a slight positive correlation between iron and tin in the copper.

Below the level of 1% the effect of tin on the properties of the alloy would be minimal, and some other reason must be sought for its presence. Tin is a powerful deoxident and it might have been added to scour dissolved oxygen and oxides from the metal. Another possible reason for its presence in small quantities might have been to promote the fusing of the prills extracted from the slag when heated together in a crucible. The melting would be considerably helped by adding a little tin which melts at $232^{\circ}C$ and into which the copper would start to dissolve. Tin is still occasionally used in this way in small specialist foundries to encourage the rapid melting of the copper. However these hypothetical uses at Timna are not supported by the discovery by the author of a 9 gram copper prill from the slag, together with a small 80 gram bun ingot from site T3, contemporary with and adjacent to the main smelting area. The bun ingot was definately cast in a crucible as part of it still adhered to the underside of the metal. Now if the above hypothesis was correct the raw prill should contain no tin, but the ingot formed by melting together many prills would contain a little tin if the initial melting had been started as described above. However, analysis failed to reveal tin in either prills or ingot and it must be hoped that further excavation will throw more light on this problem at Timna.

Many other instances of copper with small amounts of tin are known - for example the Cape Gelidonya ingots mentioned above, contain between 0.01 and 1.0 percent of tin. Similarly recent analysis of Egyptian tools and weapons by Cowell to be published in a forthcoming British Museum catalogue, showed a small number of artifacts of copper containing under 1 percent of tin. Usually such low quantities of tin are attributed to the copper ores, even though stannites, copper-tin ores are extremely rare. This present work demonstrates beyond doubt than even such small quantities can be deliberate additions, but leaves us as far as ever from understanding why they were added.

REFERENCES

Bartura, Y. and Wurzburger, U. 1974. The Timna Copper Deposit. *Gisements stratiforms et provences cupriferes* pp. 277-285. Leige.

Cooke, S. R. B. and Aschenbrenner, S. 1975. The Occurrence of Metallic Iron in Ancient Copper. *Journal of Field Archaeology 2* pp. 261-272.

Coghlan, H. H. 1975. *Notes on the Prehistoric Metallurgy in the Old World. 2nd Edition* Pitt Rivers Museum Occasional Paper on Technology no. 4. Oxford.

Holmyard, E. J. 1957. *Alchemy* p. 23, 24. London.

Hughes, M. J., Cowell, M. R. and Craddock, P. T. 1976. Atomic Absorption techniques in Archaeology. *Archaeometry* 18. pp. 19-38.

Lupu, A. 1970. Metallurgical aspects of Chalcolithic copper workings at Timna. (Israel). *Bulletin of the Historical Metallurgy Group Vol. 4 no. 1.* pp. 21-23.

Maddin, R. and   1974.   Some Notes on the Copper Trade in the Ancient
Muhly, J.D.   Mid-East. Journal of Metals 2615 pp. 1-7.

Maddin, R. and   1976.   Metallurgical study of 7 bar ingots. Israel
Wheeler, T.S.   Exploration Journal 26 pp. 170-173.

Muhly, J.D.,   1977.   The Cape Gelidonya Shipwreck and the Bronze
Wheeler, T.S. and   Age Metals Trade in the Eastern Mediterranean.
Maddin, R.   Journal of Field Archaeology 4 pp. 353-362.

Rackham, H.   1958.   Translator of Pliny, The Natural History Book
                               XXXIV. (Loeb edition) London.

Rothenberg, B.   1972.   Timna. London.

Rothenberg, B.   1978.   Chalcolithic Copper Smelting. I. A. M. S.
                               Monograph no. 1. London.

Slatkine, A.   1961.   Nodules cuprifères du Neguev Meridional (Israel).
                               Bulletin of the Research Council of Israel 10
                               pp. 292-297.

Tylecote, R.F.,   1967.   A Study of Early Copper Smelting and Working
Lupu, A. and   Sites in Israel. Journal of the Institute of
Rothenberg, B.   Metals vol. 95 pp. 235-243.

Tylecote, R.F.   1977.   Summary of results of experimental work on
                               Early Copper Smelting, in Aspects of Early
                               Metallurgy edited by W. A. Oddy. London
                               pp. 5-12.

Tylecote, R.F.,   1977.   Partitioning of Trace Elements Between the
Ghaznavi, H.A. and   Ores Fluxes, Slags and Metal during the
Boydell, P.J.   Smelting of Copper. Journal of Archaeological
                               Science 4 pp. 305-333.

Tylecote, R.F. and   1978.   Experiments on Copper Smelting in Chalcolithic
Boydell, P.J.   Copper Smelting with B Rothenberg. IAMS
                               Monograph no. 1. London.

Plate 1:   Spill of copper in the shape of a quadruped (T 200 Cat. 4)
             Cast figurine of a goat.   (T 200 Cat. 6)